ENGINEERING STRATEGIES FOR GREENHOUSE GAS MITIGATION

Controlling the level of greenhouse gas in the atmosphere is a rapidly growing area of commercial activity. It is predicted by Lord Stern in *The Economics of Climate Change* that business generated by climate change issues will rise in value to US$500 billion per year by 2050. While debate continues both about the impact of greenhouse gas on climate and the role humans play in influencing its concentration, engineers are faced with less controversial questions of how to manage this uncertainty and how to control greenhouse gases at a minimum cost to society.

This book gives a concise review of current knowledge required for engineers to develop strategies to help us manage and adapt to climate change. It has been developed from the author's graduate course in environmental engineering at the University of Sydney and Sun Yat Sen University, China. The book is written without technical jargon so as to be accessible to a wide range of students and policymakers who do not necessarily have scientific or engineering backgrounds. Appendices allow readers to calculate for themselves the impact of the various engineering strategies on greenhouse gas mitigation. The book contains student exercises and references for further reading.

PROFESSOR IAN S. F. JONES is Director of the Ocean Technology Group at the University of Sydney, Australia. He is also a director of Earth, Ocean and Space, a Sydney based environmental consultancy. He has been a visiting professor at Tokyo, Copenhagen, Concepción, Sun Yat Sen and Columbia Universities. Dr Jones holds a number of patents on greenhouse gas abatement and has lectured at many international meetings. He is the co-author of 3 books and 90 research papers, and sits on a number of journal editorial boards. He was elected a Fellow of the Institution of Engineers, Australia, and is a Councillor of the Engineering Committee for Oceanic Resources.

ENGINEERING STRATEGIES FOR GREENHOUSE GAS MITIGATION

IAN S. F. JONES
University of Sydney

CAMBRIDGE
UNIVERSITY PRESS

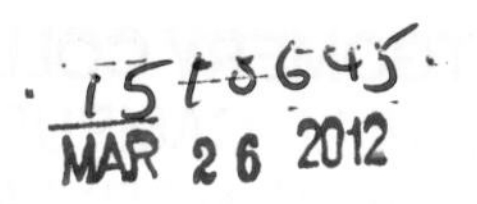

CAMBRIDGE UNIVERSITY PRESS
Cambridge, New York, Melbourne, Madrid, Cape Town,
Singapore, São Paulo, Delhi, Tokyo, Mexico City

Cambridge University Press
The Edinburgh Building, Cambridge CB2 8RU, UK

Published in the United States of America by Cambridge University Press, New York

www.cambridge.org
Information on this title: www.cambridge.org/9780521516020

First published 2011

Printed in the United Kingdom at the University Press, Cambridge

A catalogue record for this publication is available from the British Library

Library of Congress Cataloguing in Publication data
Jones, Ian (Ian S. F.)
Engineering strategies for greenhouse gas mitigation/Ian S. F. Jones.
p. cm.
Includes bibliographical references and index.
ISBN 978-0-521-51602-0 (hardback)
1. Greenhouse gas mitigation. I. Title.
TD885.5.G73J66 2011
628.5′3–dc22 2011005951

ISBN 978-0-521-51602-0 Hardback
ISBN 978-0-521-73159-1 Paperback

Contents

Preface

Controlling the level of carbon dioxide in the atmosphere is a rapidly growing new commercial activity that did not exist a decade ago. It is predicted by Stern (2007) to rise in value to US$500 000 000 000 per year by 2050. This new activity is founded on the recognition that the threat of rapid climate change is a concern for future generations. Engineers are needed to exercise their skills to deliver economic solutions to this pressing problem. Greenhouse gases such as carbon dioxide trap heat in the atmosphere and their increasing levels threaten to bring about climate change. This is a global issue and its consequences are long term. At the same time, there is much uncertainty associated with a phenomenon that is not yet understood well enough to be reliably modelled.

Last century there was much political discussion on this topic, which culminated in the agreed text of the UN Framework Convention on Climate Change (UNFCCC). With the UNFCCC entering into force in 1994, the control of greenhouse gas concentrations in the atmosphere became an engineering problem. While debate continues both about the impact of greenhouse gas on climate and the role humans play in influencing its concentration, the engineer is faced with the less controversial questions of how to manage the uncertainty and how to control greenhouse gases at the least cost to society. The modern engineer must address the concerns of the populace and will need to engage with the economist and the social scientist. Climate change raises ethical issues of intergenerational equity and the responsibility of one region for actions that damage another region. The moral issues are sharper if a damaged region is already disadvantaged. Management of global climate is an exciting new field for engineers that was considered impossible until recent times.

The root causes of the climate problem are the rising world population and the demands for an improved standard of living. There are difficulties in expanding economical food production fast enough to cope with the demand of an increasing world population. Energy is needed to grow food, and in the developed world this energy is increasingly supplied by fossil fuel. Rising gross domestic product per person is underpinned by energy consumption. As the population and living standards of developing regions rise, the consumption of fossil fuel is expected to soar. The waste product of burning fossil fuel, carbon dioxide, which is presently dumped in the atmosphere, is changing the climate and threatens to unsettle food production. At the same time more forests are cleared for agriculture and lumber, releasing still more carbon dioxide into the atmosphere. Agriculture also contributes greenhouse gases in the form of methane emissions from domesticated animals and nitrous oxide from fertiliser use. Population, climate change and food security are all connected in a way that will be addressed in the following pages.

Many fear a rapid change in climate. The international agreement signed at Kyoto in 1997 was the first step in limiting emissions of greenhouse gases. The Kyoto Protocol (see Appendix 3) is seen as more than just an agreement on managing climate change. Some see it as a move against the self-interest inherent in global capitalism. With climate change there will come a redistribution of wealth between regions, brought about both by the changes in agricultural productivity and by the demand on resources caused by mitigation efforts. Regrettably there is a reluctance to invest in addressing the future problem of further climate change. The Kyoto Protocol is different from earlier agreements such as the Patent Cooperation Treaty. Patents confer economic rights on one group, possibly to the disadvantage of another group of people. The patents on drugs to treat the AIDS pandemic dictate costs which may exclude impoverished sufferers from gaining treatment, to the disadvantage of all humankind. The Kyoto Protocol, in contrast, asks countries to forego economic benefits, to the advantage of all humankind.

Debate has continued as to whether the changes in climate, exemplified by the warming of the earth over the last few decades, are due to ‘natural’ or anthropogenic causes. The consequences of burning fossil fuel open up a whole new field of social justice, as the issues of who will benefit and who will pay are sorted out. If we are looking for someone to blame and make financially liable for the disadvantages of change, the distinction between natural or artificial is important. ‘Natural changes’ can be considered to be no one’s fault! As we are speaking of managing ‘global commons’, the usual national structures that provide some measure of social justice within a country do not apply in this international situation. New concepts will be needed.

Anthropogenic climate change raises issues of environmental ethics. By carelessly disposing of waste today, we leave a problem for future generations. The long-term consequences invite discussion of intergenerational equity. There is the problem of predicting the future; of recognising when to take action in a gradually transforming climate; of providing insurance for dramatic change such as melting of the ice caps, the rerouting of ocean currents or the warming of methane hydrate deposits. Then there are those who say the climate is too complicated to control. This is just an assertion. These possibilities have different probabilities of occurrence and different severities of consequence.

The study of greenhouse gas mitigation is a new discipline with many different facets. The burning of fossil fuels for energy and the clearing of land for agriculture have been the keystones of progress that has lifted ordinary people in the developed countries out of material poverty and want. Now we find these practices may be changing the ecology of our surroundings in an unsettling way.

Engineers must make decisions that recognise the 20- or 30-year lifetime of large projects. They do this within a policy framework set out by others. They have an increasing need for the support of social scientists in identifying communities' concerns about innovation. Analysing the risks of climate change and designing the mitigation strategies requires a complex balancing of financial, ethical and economic issues. This must be done in the presence of much uncertainty, not just about the physics of climate change and its cause, but about the future economic progress of humankind. How much should engineers worry about possible climate change that flows from their design? The rules of how the engineer should act in the face of such uncertainty must be developed. What are his or her options? This book attempts to set out the strategies available and to examine in detail the engineering issues.

Already there is much material available on the science of greenhouse gas, but little discussion on mitigating the change in global climate. While we have a call to action in the Kyoto Protocol, there is not a good synthesis of the literature available on choices. This book hopes to fill the gap by reviewing current knowledge needed to allow engineers to manage climate change. The companion activity of adapting to the impacts of rapid climate change is also touched upon. In many cases this may be the more efficient allocation of resources.

This text has been used as the basis of a graduate course in environmental engineering at the University of Sydney for a number of years.

Acknowledgments

I saw my mother as a role model. My colleague, Helen Young, discussed these ideas with me for a decade and remained patient as she dissuaded me from the more extreme and ill considered.

1

The future greenhouse gas production

Introduction

Carbon has been stored in the trees, vegetation and soil of the earth and as fossil carbon in the forms of coal gas and oil for millions of years. The Industrial Revolution of the eighteenth century, combined with expanding land use for agriculture to feed the rapidly rising population, has transferred significant amounts of this carbon to the atmosphere. The rising affluence of an increasing global population is predicted to lead to the release of more greenhouse gas to the atmosphere in the future. With industrialisation's present dependence on fossil fuels for economic energy, there is unlikely to be restraint in the use of fossil carbon. The generation of greenhouse gas is expected to rise, particularly in rapidly developing countries such as China and India. Concentrations of these greenhouse gases in the atmosphere will increase unless there is deliberate expenditure on enhancing carbon sinks. With this change in concentration of greenhouse gas will come a change in the climate. The central theme of this book is how to have more economic growth using low-cost energy and at the same time sustain the earth's environmental quality. Engineers need to think in the time frame of 50 years because the consequences of their decisions on capital investments will be with us for many decades. To provide a better future, engineers need to constantly embrace innovation and plan for the environmental change that results from their actions. The management of the earth's environment can no longer be left to nature. Nor should all environmental change be assumed to be undesirable. Some change is now unavoidable, and society will need to adapt.

It is only recently that there has been the realisation that human activity has the potential to influence global climate, and it is only recently that the human population has been in large enough plague proportions to do so. The English economist Thomas Malthus was the first to raise our consciousness that the rising population was a danger to humankind's wellbeing. When he pointed out the exponential

Figure 1.1 S. Arrhenius (1859–1927), a Swedish scientist who predicted global warming.

nature of population growth at the start of the nineteenth century, he was roundly criticised by exponents of limitless growth, who argued that with growth came the prospect of social equity. However, this goal is yet to be achieved, while the world population has approached a crisis level.

The Industrial Revolution which occurred in some parts of the earth was fuelled by plentiful fossil fuel that could be extracted cheaply and transported easily. Arrhenius (1896; Figure 1.1) realised that the carbon dioxide from the burning of fossil fuel could lead to a change in the earth's temperature. He predicted an increase of 5–6 °C with a doubling of the concentration of carbon dioxide. However, there was general scepticism and Weart (1997) described the long period of resistance by scientists to the concept of global warming as a serious possibility.

Before we can assess the threat of climate change to future generations we must make some predictions about the future. This chapter presents a number of possible scenarios for the next 100 years, and sets the stage for looking at techniques for mitigating the increase of atmospheric carbon dioxide.

The natural carbon cycle

The world runs on a carbon cycle. Animals burn carbon by breathing in oxygen and exhaling carbon dioxide. Plants, through photosynthesis, absorb this carbon dioxide and return oxygen to the atmosphere. In the ocean, phytoplankton incorporate carbon dioxide and release oxygen. Fish breathe the oxygen and feed on the phytoplankton. This carbon cycle has been in approximate equilibrium for thousands of years.

Since the Industrial Revolution (of the mid-eighteenth century), the carbon cycle has been out of balance, with the carbon dioxide level in the atmosphere rising steadily. It is the burning of fossil fuels and the clearing of forestlands for agriculture by humans that has caused much of this anthropogenic change in carbon dioxide levels.

It is clear from Figure 1.2 that for many centuries the partial pressure of carbon dioxide in the atmosphere was near 280 parts per million by volume. The observations for the earlier years were obtained by analysing gas bubbles trapped in cores taken from ice shelves. The values for the last 50 years are from direct measurements in the atmosphere, and the values for the next 100 years are the predictions of the carbon dioxide concentration under a number of models of the future. We will discuss the scenarios used in these models below. They are word pictures of the future grouped into families with similar storylines. Nakicenovic and Swart (2000) designated the main storylines as A, B, C and so on, and variations within the storyline as 1, 2, 3 etc.

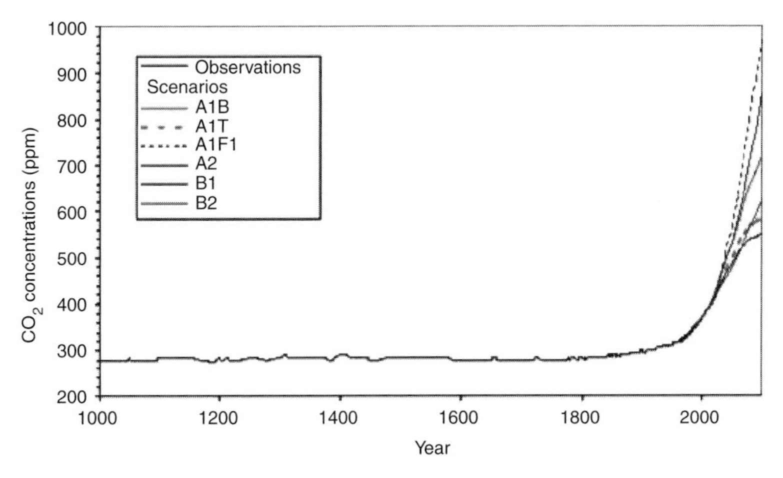

Figure 1.2 See plate section for colour version. The change in atmospheric concentration of CO_2 over the last thousand years. Various future scenarios are also shown. The A2 scenario is described in more detail in the text; for others refer to Nakicenovic and Swart (2000). Note the truncated vertical axis. (Is this a good engineering practice or is it a technique to raise alarm?) Reproduced from Houghton *et al.* (2001).

The A2 storyline and scenario family, for example, describes a very heterogeneous world. The underlying theme is self-reliance and the preservation of existing local customs. High population growth persists in some regions, which results in a continuously increasing global population. Economic development is primarily regionally oriented, and per capita economic growth and technological change are more fragmented and slower than in some other storylines. The A2 scenario is based on a high population growth to 15 billion people by the year 2100.

From these visions it is necessary to predict the changes in gross domestic product (GDP), percentage population change and carbon intensity in the future. With such figures and the aid of Equation (1.1) below, the predictions of carbon dioxide concentration in the atmosphere which are shown in Figure 1.2 can be constructed.

We can see that the predictions of the partial pressure of carbon dioxide, pCO_2, in the atmosphere for the various scenarios illustrated in Figure 1.2 are all similar, but we should not take comfort from this as most 'reasonable scenarios' assume the world social order will continue much as at present with about 1000 million affluent people, some poor but contented and a few billion desperately poor people. Such a situation may not be sustainable now that television provides opportunities for the disadvantaged to know their plight and for them to engage in terrorism to redress the perceived injustice.

In order to make reasonable predictions of the future we must face these all-embracing questions of the nature of society, the distribution of wealth and the political ideologies that triumph.

What is greenhouse gas?

A number of atmospheric gases are responsible for trapping solar radiation in the atmosphere. The radiation from the sun, which is very hot, is of short wavelength; while the radiation into space from the earth, which has a surface temperature of about 15 °C, is of much longer wavelength. Atmospheric gases absorb the long-wave radiation that would, in their absence, escape from the earth and radiate into space. It is the change in radiation forcing of the earth caused by these gases that will drive a variation in the climate. Although water vapour is the most important greenhouse gas (it absorbs the most radiation) it is not usually considered in discussions of climate change forcing but rather is considered part of the hydrosphere response to changing radiation. Why this is so is not clear. With an increase of global temperature is expected to come an increase of water vapour in the atmosphere. This is a positive feedback and may double the global warming brought about by increases of other greenhouse gases.

Table 1.1. *Direct global warming potentials. Relative forcing per kg of trace gas. These direct effects do not include indirect effects due to interactions involving atmospheric chemical processes.*

Trace gas	20-year time horizon	100-year time horizon
CO_2	1	1
CH_4	62	23
N_2O	275	296
SF_6	15 100	22 200
HCFC-23	9 400	12 000

Source: Houghton *et al.* (2001).

Houghton *et al.* (1990) show the relative influences on radiation forcing (in the absence of feedback) of a number of trace gases, and these are reproduced in Table 1.1. Notice that relative forcing depends upon the lifetime of the gas in the atmosphere.

A simple measure for estimating the climatic effects of greenhouse gas is the change in radiative energy flux on the earth. Global warming potential per unit of mass is the change in radiative energy flux due to one kilogram of gas relative to the change due to one kilogram of CO_2 released into the atmosphere. Note that CO_2 warming potential also changes with time, as a fraction of it enters the ocean. Concern has been focused on carbon dioxide and nitrous oxide, N_2O, as they have a relatively long lifetime in the atmosphere, while the lifetime of methane, CH_4, is much shorter. Table 1.1 shows the relative greenhouse gas forcing of methane decreasing with time.

Clouds are droplets of water formed from water vapour that evaporated mostly from the ocean. Two-thirds of the surface of the earth is ocean. Clouds both trap long-wave radiation and reflect short-wave solar energy, as well as influencing rainfall. They can either cool the planet or lead to global warming. This represents a significant source of uncertainty. The impact of changing cloud cover will need to be better understood if the prediction of the future climate is to be made more reliable.

No further serious discussion is made of gases other than CO_2 in this book, since CO_2 is predicted (in a *business as usual* strategy) to dominate the change in greenhouse gas concentration over the next 100 years. An exception to this is the role that agriculture plays in the continual conversion of carbon to methane. Increasing agricultural activity to feed the extra billions of people predicted to live on earth will, in the absence of changed practice, increase the emission of CH_4.

What is climate change?

Measures of the environment (such as air temperature) fluctuate from time to time and from place to place in what seems a random manner. By climate we mean the statistical measures of meteorological and oceanographical variables such as temperature or wind speed. Climate can be thought of as a group of statistics such as the mean temperature or the probability distribution of wind speed when many days of record have been considered. The weather, on the other hand, is the descriptions of the amount of rainfall, for example, over a much shorter timescale, such as a day. The climate is a measure describing the 'normal' weather. The seasonal change in temperature at a point, for example, can be described as a conditional average (that is, conditional on the season).

Time series of variables, such as the daily mean temperature, exhibit seasonal, yearly, even century-long fluctuations. There are fluctuations in statistical measures of the environment on all timescales and so it is not useful to think of a steady climate. 'Is the climate changing?' is not a question we can answer. 'Has the climate changed?' The answer to this question is that climate based on decadal averages has been changing.

One cannot make categorical claims using statistical theory about whether statistical variables will change in the future. Statistical data describe the past. Under the assumption that the statistical behaviour of the variable is steady, the future can be predicted. One can make more complicated models, based on assumptions such as that the rate of change of a variable is constant. Such assumptions are reasonable for short-term predictions but most unreliable for long-term (100-year) estimates. Physical models, on the other hand, can predict the future but are imprecise because our assumptions are based on limited understanding of the physical processes. One can report on the past, but the future remains uncertain! While this may represent a problem to the politician or the journalist, engineers are used to proceeding in the face of uncertainty. Managing risk is the hallmark of an engineer!

Some scenarios place the changes in radiative forcing (see page 19) due to CO_2 over 50 years at 3 W m^{-2} (to be compared with solar radiation of 340 W m^{-2} at the top of the atmosphere, averaged over the earth). This might lead to a rapid climate change of 0.5 °C per decade. Sea level, rainfall patterns and frequencies of severe storms all will change. Will the change be good or bad? We will return to this question.

As we burn fossil fuel, we release heat as well as carbon dioxide. The solar radiation that heats the earth is about 120×10^{15} watts, while humans produce only 0.018×10^{15} watts by burning fossil fuel. Energy obtained from the sun, the renewable power source, does not heat the earth. Thus climate change is unlikely to be driven by waste heat.

While many people focus on the change induced by humans, that is, anthropogenic change, the consequences of climate change are unrelated to the cause. The United Nations Framework Convention on Climate Change uses the word differently to this book. It refers to *climate change* as a change of climate that is attributed directly or indirectly to human activity that alters the composition of the global atmosphere. This is anthropogenic change. It may be that the Convention has adopted this peculiar definition because it is our social custom to expect people who induce change to pay for the negative consequences. For instance, when engineering works damage adjacent buildings, the owners of these buildings expect compensation.

Sources of carbon dioxide

Each person contributes to the emission of carbon dioxide into the atmosphere by breathing, burning firewood, clearing forests, using fossil-fuel-fired electricity and by driving motor cars or burning sugar cane. Some of these activities make no medium-term change in the atmospheric concentration of carbon dioxide. The burning of firewood, for example, makes no long-term net change to atmospheric carbon dioxide if the wood is regrown, because the carbon that is released is extracted from the atmosphere in the new growth of the trees. The burning of fossil fuel such as coal for electricity generation, however, is different. Here carbon has been locked away for millions of years and is suddenly released to the atmosphere. Cars and aeroplanes running on petroleum are also burning fossil fuel.

While this process of activating fossil carbon involves some six gigatonnes per year of carbon, it is only a small fraction of the natural carbon cycle. Over short times, carbon exchange between the atmosphere, the ocean and the land is at the rate of hundreds of billions of tonnes per year. Why is there so much concern about such an apparently small perturbation? It is that the large fluxes in and out of the atmosphere are nearly in balance. The flux of carbon into the ocean, over areas of hundreds of km^2, and averaged over periods of months, is about 2–4 $GtC\,yr^{-1}$, and out of the ocean it is 1–2 $GtC\,yr^{-1}$. That is to say, some regions of the oceans are degassing and some are absorbing carbon dioxide from the atmosphere. Averaged over the seasons there is a net sequestration of order 2 $GtC\,yr^{-1}$ into the ocean as a whole, and this comes about, as we shall see in Chapter 5, because the concentration of carbon dioxide in the atmosphere is rising. Fluctuations in the ocean circulation will lead to yearly changes in the net uptake, and these can be seen as variations in the yearly march of carbon dioxide concentration in the atmosphere. Into the land biosphere there is another flow that varies seasonally and is about 2 $GtC\,yr^{-1}$ averaged over the year. It is the remaining two billion tonnes per year of carbon that is causing the change in partial pressure in the atmosphere that can be seen in Figure 1.2. This net emission is slowly changing the total carbon dioxide in the atmosphere. At present the atmosphere contains about 750 Gt of carbon in the form of carbon dioxide.

Estimating carbon dioxide release to the atmosphere

We can estimate the amount of greenhouse gas produced by each person and add this up for everyone on the planet. The production of greenhouse gas per person can be predicted from the gross national product (GNP) per year and per person, *GNPp*, multiplied by the greenhouse gas intensity of the region in which that person lives. Greenhouse gas intensity, I, is a measure based on the equivalent amount of anthropogenic carbon dioxide released for each unit of GNP. In the USA, for example, the intensity is about 180 tonnes of carbon per million US dollars of GDP (GNP and GDP will be used interchangeably as their differences are not significant in the present approach). Equivalently this is 0.7 $kgCO_2$ $\$^{-1}$ and has been falling for the last 50 years as engineers learned how to use energy more efficiently. Note we are only considering emissions that are a result of manufacture or farming that do not have an associated sink.

Then it is just a matter of multiplying the greenhouse gas per person by the number of people, P, to produce the expression

$$\textit{Anthropogenic carbon dioxide emission} = P \times GNPp \times I \tag{1.1}$$

When the average intensity over the whole world population is used, the present emissions can easily be estimated. Note, we are talking of fossil fuel emissions and emissions from land change practices, not emissions from breathing or burning firewood. Often people talk of carbon dioxide equivalent emissions. This is where an allowance for the emission of other greenhouse gases that increase with GDP is also included with a weighting depending on their radiation forcing potential. It is also important to distinguish between the emission of greenhouse gas – Equation (1.1) – and the *net* anthropogenic emission which is the difference between the emission of anthropogenic carbon dioxide equivalent and the artificial sinks of carbon dioxide.

Since the ocean absorbs more carbon dioxide than it releases back to the atmosphere, it is a sink of carbon at present. However the present uptake is not due to any intentional action of humans and so is not considered to reduce the net emissions. Land-based plants, under increasing atmospheric concentrations of carbon dioxide, are also sinks of carbon, as they take carbon from the atmosphere to produce their woody biomass. These can be thought of as feedback mechanisms and occur without human intervention. This distinction between natural and anthropogenic can become a little confusing. One assumption is that the natural sources and sinks of carbon dioxide are in balance, but the evidence is mainly that the concentration of carbon dioxide in the atmosphere before the Industrial Revolution (Figure 1.2) was constant. One cannot be sure about the last 150 years. Under the assumption of zero 'natural net emissions', the increase of carbon in the atmosphere can be obtained by subtracting, from the anthropogenic emissions of carbon dioxide, the sinks of greenhouse gas (that are feedbacks or are deliberately human induced). The latter are anthropogenic sinks such as the conversion of farmland to forest that will be discussed in later chapters.

Knowing the current anthropogenic emission level is useful, but there is a greater need to be able to predict the future emissions if we are to take appropriate action now. If the net amounts were to soon return to zero, as they were a few hundred years ago, there would not seem to be a long-term problem of greenhouse-gas-induced climate change to address. If the emissions were to soar, the threat of rapid climate change would be large. Predicting the future is not easy. Nor is it at all certain. To try and predict future greenhouse gas emissions we need a scenario of the future population, wealth of the globe and carbon intensity. The level of wealth will vary according to how the future unfolds. Thus we have prepared a number of scenarios of the future, as have others such as the Inter-Govermental Panel on Climate Change, e.g. Nakicenovic and Swart (2000).

The scenarios below are not predictions but rather visions of the future that try to embrace a range of possibilities. There is no likely future! How much was the end of the twentieth century like the end of the nineteenth century? Could you have predicted the increase of GDP of the Western countries? One popular way of thinking is to imagine the future as an extrapolation of the past, but there have been revolutions that have shown this approach is flawed. It is generally unsettling for those with a comfortable lifestyle to think there will be a revolution. By revolution people mean a break from the past. We must accept that the future may be wildly different to any of the five scenarios below.

Scenarios of the future

Humankind lives in harmony with nature – requires the population increase to be controlled and quickly stabilised, probably reduced; impoverished communities will obtain better nutrition, health, education, peace and stability; rich communities will exhibit less materialism, more social conscience and there will be more even distribution of existing resources; a change in value systems within consumer societies will occur, so that consumption per person is reduced. More organic farming will be practised as the demand for food reduces with population.

Business as usual – global capitalism spreads over the world in the control of people in a few sovereign states; while multi nationals seek the cheapest labour and the highest prices for their goods and services, they repatriate their profits to a few nations; land clearing for cities and agriculture continues and non-sustainable farming tries to feed increasing populations but still the number of hungry rises. An elite exchange information on the internet while billions remain uneducated and burn the cheapest remaining fuel for cooking and warmth – fossil fuel consumption rises steadily. Debates about the importance of greenhouse gases continue but no further action on mitigation occurs after 2012 when the Kyoto Protocol expires (see Appendix 3).

Anarchy – survival of the fittest – the elite live in enclaves, protected by force and controlling the impoverished majority by restriction of their movement and provision of their goods and services as in former apartheid Africa. There is little thought for the future because the threat of anarchy makes survival uncertain. Research and development decline. Energy use rises little, but fossil fuel is used inefficiently by the poor. Poverty increases mortality through diseases such as HIV Aids.

Brave new technological world – education eliminates racism, technology provides ever cheaper health and replacement organs for humans so mortality drops; distance learning provides universal literacy; humankind discovers renewable, sustainable energy at minimal cost; the elite of the new society insist on equal opportunity; population numbers decline because fecundity decreases voluntarily since support for the elderly is guaranteed by the community rather than by family. Fossil fuels are no longer burned and agriculture can be practised on less land.

Ludditism – a foolhardy non-interventionist philosophy (risk averse) which encourages a 'do nothing' approach to prevail; e.g. society bans genetic food and information about birth control; there is a return to 'fundamentalism'. Conventional farming leads to a decline in food production per acre as soil becomes impoverished; natural gas for fertiliser production becomes scarce. Poverty becomes more widespread. Fossil fuel is burned at the present level of efficiency. Populations rise until mortality due to poverty exceeds fertility.

All the above will induce a change in the level of greenhouse gas in the atmosphere as the fossil fuel burning and land use practices will change. Land use modifications will influence the fraction of solar energy reflected from the earth, called the earth's albedo.

Predictions of emissions

With each one of these storylines, one can make estimates of what are appropriate values for population growth and so on. Some storylines imply continued population growth, while some imply an excess of mortality over fertility. There are many other scenarios, such as a large global recession following the unprecedented growth of the first decade of the century to bring the long-term global growth of GDP back in line with the previous century.

One of the earliest *business as usual* scenarios, presented by the Inter-Govermental Panel on Climate Change, was known as IS92b. In this scenario, Houghton *et al.* (1992) predict the changes in terms of Equation (1.1) if current trends are continued without major change in values of the people. You will find us returning to this scenario throughout the book. It assumes that the population reaches 11.3 billion by 2100, that global economic growth is between 2.9% and 2.3% per annum and that there is compliance with the Montreal Protocol. (The Montreal Protocol restricts the emission of gases that reduce the ozone level in the atmosphere and produce what is popularly known as the ozone hole.)

Business as usual models

To determine the global emissions of carbon dioxide released into the atmosphere from anthropogenic causes, we will divide the world into the 13 regions shown in Figure 1.3. Then the changes in each region can be treated separately and consolidated into a world picture.

Present predictions of population are based on estimations of the future fertility rate minus the mortality. When regions are considered separately, the migration of people also needs to be estimated but of course the net migration for the world is zero. Figure 1.4 shows the predicted change in population over the

1 Canada
2 USA
3 Latin America
4 Africa
5 OECD Europe
6 Eastern Europe
7 CIS
8 Middle East
9 India and South Asia
10 China and CP countries
11 East Asia
12 Oceania
13 Japan

Figure 1.3 The thirteen regions of the world for scenario planning, following Alcamo *et al.* (1994a).

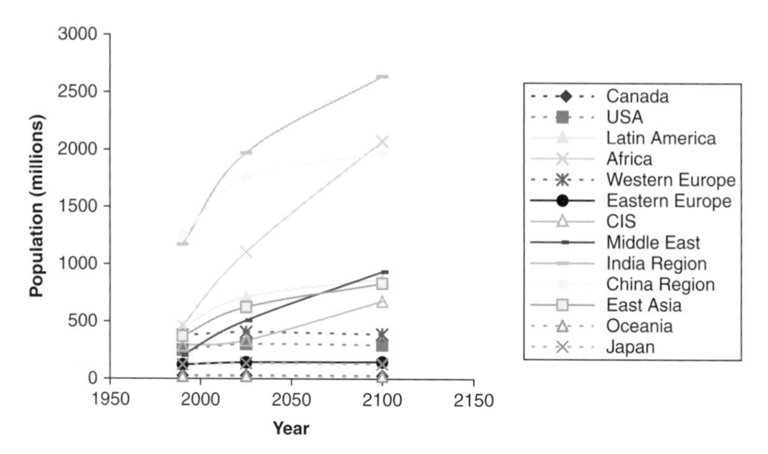

Figure 1.4 See plate section for colour version. Population prediction that Alcamo *et al.* (1994b) used in IMAGE2 for the regions shown in Figure 1.3.

next 100 years in the 13 regions of the world (Figure 1.3) that were used in the IMAGE2 model of Alcamo *et al.* (1994a). The operators of the model suggest that the population will be reasonably steady in developed countries but rising rapidly in Africa and the Middle East, see Table 1.2. These estimates are consistent with those by the World Bank and UN demographers who have tried to include the impact of disease and increased longevity. In the past it has been found that the population increase has been small in regions where the GNP per person is high.

The gross domestic product is some measure of the size of the domestic economy of a region. There are many difficulties in using this measure as it does not include non-money transactions. Unpaid domestic work is not included, nor is the food consumed in subsistence farming. When annual changes in GDP are applied for many years, the total change can be spectacular. Table 1.3 shows that a growth rate of 3% per annum would lead to a 25-fold increase in GDP after 100 years. Such big changes cannot be imagined by many people, which leads them to ridicule such figures. In 1990 developing countries had only 20% of the total GNP, despite representing 80% of the global population.

We also wish to note that we will try to express all quantities in terms of dollars of constant purchasing power. This is not easy to do and often the current exchange rate will be used. As well, inflation changes the purchasing power of a currency and so will be excluded by expressing things in terms of US dollars at their value in the year 2000.

The predicted gross national product (GNP) per person, used in IMAGE2, is shown in Figure 1.5 for these 13 regions, with the developed countries distinguished by broken lines. Even after 100 years, in this scenario the disparity

Table 1.2. *Population in millions.*

Region	1990	2025	2050
Canada	27	29	28
USA	250	302	298
Latin America	448	715	824
Africa	642	1540	2208
OECD Europe	378	407	395
Eastern Europe	123	143	149
CIS	289	335	350
Middle East	203	508	730
India and South Asia	1171	1970	2375
China and CP countries	1248	1756	1896
East Asia	371	624	750
Oceania	23	25	24
Japan	124	136	132
World	5297	8490	10161

Table 1.3. *Compound interest tables.*

Annual percentage change	Change 1990–2100
2	8
3	25
4	74
5	214
8	4700

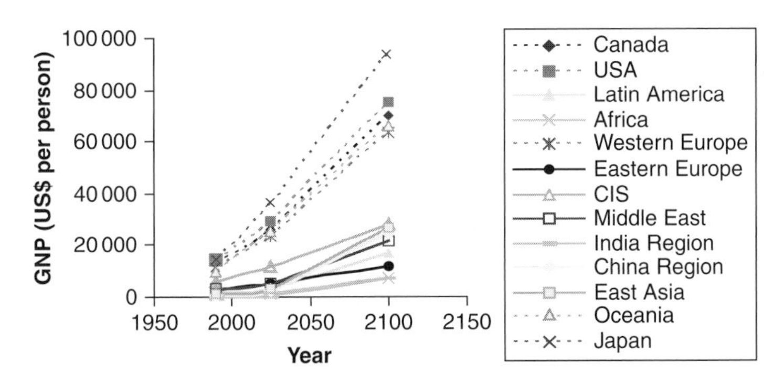

Figure 1.5 See plate section for colour version. Gross national product per person in 13 regions. By the year 2100, Japanese residents are predicted to have an average income of US$95 000. After Alcamo *et al.* (1994b).

between the industrialised and the agrarian countries remains large. The prediction of increase of GNP is uncertain because it makes assumptions about the success of political and economic systems. Such judgements are often influenced by ideology. The estimates used in IMAGE2 are an extrapolation from immediate past performance, that is, *business as usual*. The figures show the average resident of Japan to be nine-times as wealthy as a resident of the Indian region in the year 2100. Will large populations accept this disparity? Will the disparity of GNP per person between regions persist for the next 100 years? Will there be revolution? Will there be a major war or terrorism activity like that which brought to an end the Western Roman Empire in 505 CE? Will there be a shift in thinking similar to the complete shift in the acceptability of keeping slaves that occurred in the nineteenth century? It is possible that it will be considered unacceptable that millions of people live in poverty and misery while a small elite consume conspicuously.

Figure 1.5 makes no comment on the distribution of income within the region. In the recent past the increase in GDP per person has not meant that the poor have access to greater income. It meant that the rich became richer.

Some people consider these predictions ridiculous. Some think that such levels of consumption are unsustainable. If we think back 100 years to Western Europe at the turn of the century, about 10% of the population lived in comfort while 90% struggled to obtain education and health services. Now, at the start of the twenty-first century, 90% of Western Europeans live comfortable lives and only 10% struggle. In 1900 very few people would have predicted this transformation of the bulk of Western Europeans to middle class. We can look at some historical figures of the GDP per person in Figure 1.6.

Rich countries have been able to achieve more than 2% pa growth over the last 40 years. However the poor countries have not done so well over the last 20 years. Currency units create some uncertainty, and more extensive statistics can be obtained from the World Development Indicators published by the UN World Bank.

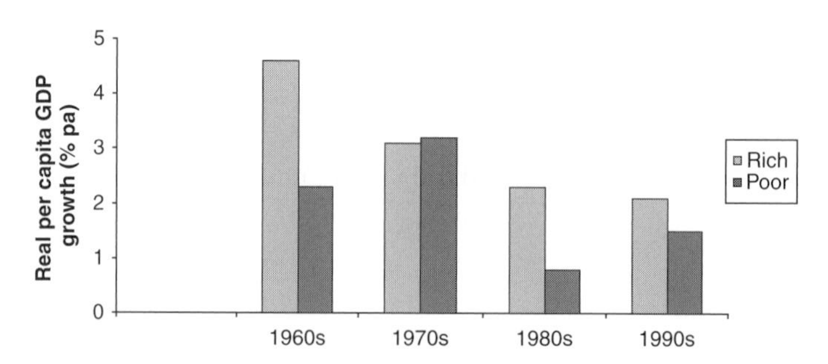

Figure 1.6 Real GDP growth for rich countries, and the average of 48 poor countries.

They suggest that China has been able to achieve economic growth of the order of 10% pa for the last two decades. Historically there have been cycles in economic growth, with periods of recession following periods of high growth such as experienced by China in the first decade of the twenty-first century.

Economics is a study of the behaviour of people with regards to money. The response of people, faced with different situations in the future, cannot be reliably predicted today. The ratio of savings to immediate expenditure can change with time. Neither can the uncertainty be reduced by the collection of more statistics. Precise knowledge of the past does not reduce the uncertainty of the future. Predictions themselves influence the measures of economic activity, as we will discuss later when we look at how the possible negative impacts of climate change have influenced countries to sign the Kyoto Protocol. Predictions of the economic costs from climate change, if the world pursues a *business as usual* approach to greenhouse gas emissions, are changing peoples' economic behaviour. When all these issues are considered it can be seen that *reliable* prediction of GDP in 100 years is not possible at present. However, almost any prediction is better than none.

Rather than assume that the future will follow the past, as in Figure 1.5, economic models can be used. The GDP per person can be predicted by computerised, dynamic, general equilibrium models of the economy such as the DICE model developed at Yale University (Nordhaus and Boyer, 1999). Such models assume a certain population trajectory and preference by humans between immediate consumption and saving for the future. It also includes estimates of economic damage as a function of greenhouse gas concentrations which in turn depend on the economic activity. With these assumptions, the model makes predictions in mathematical terms about the future.

These models suffer the same uncertainty as the more direct estimates of quantities such as future GDP. Rational economics is of little value in predicting future GDP, as people seem to vary investment and consumption for 'irrational' reasons. As well, human values can be expected to change over the 100 years that is appropriate in considering climate change. Economics is badly infected with ideology, and economic modelling of climate change is no exception. In particular, the discounting of the future value of the environment is a point of contention. Nordhaus (2007) discusses the different approaches taken to the time discount rate by the Stern review and the DICE model described above.

With predicted values of GNP or GDP per person and population we have two of the terms needed to predict anthropogenic greenhouse gas emission. Figure 1.7 shows the predicted total GDP under simple assumptions consistent with business as usual.

The assumed GDP per person growth of 2% pa can be considered in historical perspective by looking at Figure 1.5. By combining GDP and intensity change in Equation (1.1) we can obtain the emissions. Some intensity values are given in

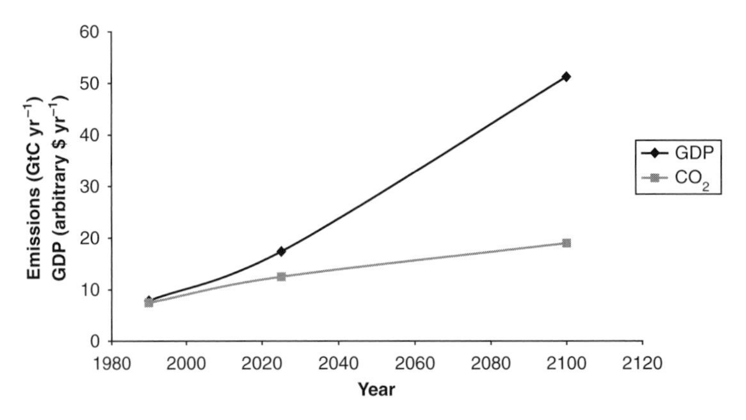

Figure 1.7 Carbon dioxide emissions in gigatonnes of carbon per year will rise more slowly than economic activity. The GDP is expressed in arbitrary values. (This curve assumes a population growth of 0.65% pa, a GDPp of 2.0% pa and a greenhouse gas intensity of −1.3% pa.)

Table 1.4. *A few figures on global emissions of carbon dioxide useful for greenhouse gas intensity estimation.*

Year	Fossil fuel and cement (GtC yr^{-1})	Global GNI index (constant $)	CO_2/GNI	Percentage change pa
1980	5.297	1.00	5.3	
1990	6.126	1.39	4.4	−1.7
2000	6.608	1.81	3.64	−1.7
2100			0.66 (estimate)	−1.7 (prediction, BaU)

BaU = business as usual.
Emissions, http://cdiac.esd.ornl.gov.
Notice we are using gross national income (GNI) rather than GDP.

Table 1.4. We have assumed in calculating the emissions that the amount of energy per unit of GNP decreases at about 1.3% pa. Without this assumption of increasing efficiency, the carbon dioxide emissions into the atmosphere would be much greater.

Prediction of concentration

The concentration of carbon dioxide in the atmosphere depends upon the mass of carbon dioxide gas that resides in the atmosphere. We know this is about 750 GtC at present. There is a continual flux of carbon (dioxide) across the atmosphere–land boundary and the air–sea interface. The sum of these fluxes determines the change in the amount of carbon in the air and consequently the concentration. We will think of three fluxes: anthropogenic emissions discussed in the above section; carbon that flows in or out of the soil, terrestrial plants and animals; and the flux into the sea. We are not thinking of the daily or weekly average fluxes but of averages over periods longer than the seasons. Uptake is positive when carbon flows to the atmosphere and it is negative for flux from the atmosphere. Emissions are positive. The amount (mass) of carbon in the atmosphere at a particular time is given by

$$Mass = 750 - ocean\ uptake - land\ uptake + emissions \tag{1.2}$$

where 750 GtC is the present mass. Note that it will change with time.

It is often assumed that, in the absence of humankind, the mass of carbon in the atmosphere would be nearly constant, or at least changing much more slowly than it is changing at present. It is assumed that, in the absence of humans, the yearly average of the ocean and land uptake would come to zero. If we wish to study the change of mass of carbon in the atmosphere due to the human-induced emissions alone, we must adjust the anthropogenic emissions by the changes that these emissions induce in the other sinks, the feedback fluxes. The increased CO_2 concentration that has come from burning fossil fuel has increased the flux of carbon dioxide into the ocean by some 4–8 $GtCO_2\ yr^{-1}$. The molecular weights of carbon and carbon dioxide are 12 and 44 respectively. If the concentration continues to rise and the wind patterns stay the same, we expect this flux to increase (become more negative). Thus we must subtract the non-anthropogenic uptakes that are feedbacks that occur without human invention to obtain the amount of anthropogenic carbon remaining in the atmosphere. When we deliberately create a sink, such as planting a tree, we call this an anthropogenic carbon sink. Thus we write

$$\begin{aligned} Change\ in\ mass\ of\ atmospheric\ C\ =\ & anthropogenic\ emissions \\ & + ocean\ sinks \\ & + land\ sinks \\ & - anthropogenic\ sinks \end{aligned} \tag{1.3}$$

Both ocean and land sinks exist because of the anthropogenic emissions of CO_2, and are the *natural* response of the ocean and the land to changing carbon concentration in the atmosphere. The ocean sink is a result of the Henry's Law response to the atmospheric change, and involves no human intervention. Henry's Law can be used to deduce that the concentration of carbon dioxide below a water surface is proportional to the partial pressure of carbon dioxide in the air above. Since concentration of CO_2 in the air has been rising since the Industrial Revolution, some of the atmospheric carbon has flowed to the sea. It is interesting to note that, given enough time, much of the fossil carbon that is released into the air will end up in the sea. It takes some thought to decide if flow of carbon in or out of the atmosphere should be counted in the pre-industrial component and so be part of the balance, or whether it should be included in Equation (1.3). Is setting a forest fire creating an anthropogenic emission? Is the carbon uptake as a result of planting trees after clear felling of a forest creating an anthropogenic sink?

The concentration of carbon dioxide can be calculated from the mass of carbon using the gas laws. Note that concentration is usually expressed as parts per million by volume, while emissions or sinks are measured in tonnes per year.

The 1990 concentration of CO_2 in the atmosphere was 370 ppm by volume. One of the difficulties in predicting the change in carbon dioxide concentration in the future is that of calculating the *natural* uptake by the land and ocean. Some of the change results from the response of these systems to carbon dioxide fertilisation, and some to consequences of unintentional human activity such as agricultural practice or the run off of fertilisers from agricultural activities to the ocean. With rising anthropogenic emissions (and with no increase in the sinks), the partial pressure of carbon dioxide in the atmosphere will rise rapidly. If the *business almost as usual* emission scenario of IS92a is followed, the atmospheric concentration of carbon dioxide in the year 2100 is expected to exceed 750 ppm. We can only make this assertion with the help of a model of the changing nature of the fluxes to the land and the sea. It implies that the 'natural' response of the land and ocean systems cannot keep up with emissions. If we wish to restrict the rise in CO_2 to 550 ppm in the year 2100, net emissions would have to be substantially less than that of scenario IS92b. By net emissions we mean the anthropogenic emissions minus the anthropogenic sinks. Figure 1.8 shows the concentrations of carbon dioxide that

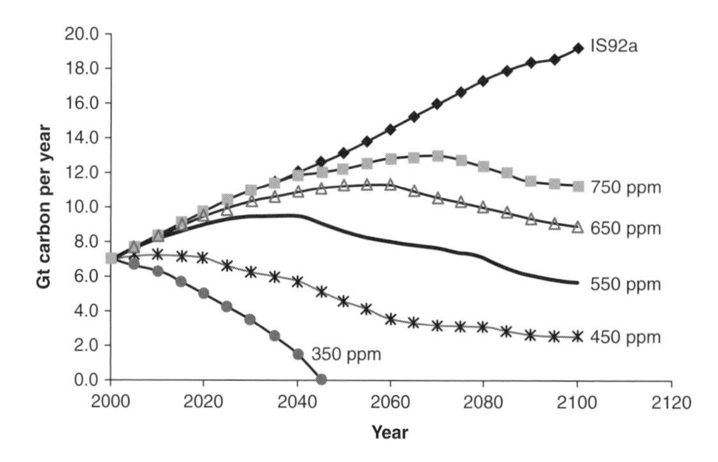

Figure 1.8 Annual emissions after Wigley *et al.* (1996). Notice that the carbon dioxide concentrations shown are those that would be achieved by about 2100, by the history of net emissions per year shown.

would be achieved in 2100 if emissions were to change with time as shown in this figure. Current emissions in the year 2000 were about 28 $GtCO_2$ (7 Gt carbon). If we wish to achieve a concentration of 550 ppm in 2100, net emissions could rise for the next 40 years and then they need to fall back to current values by 2080 and finally drop to 5.5 $GtC\,yr^{-1}$ in the year 2100. This is a big reduction from IS92b, where it is assumed that an increasing fraction of people adopt a 'Western lifestyle' with 'profligate consumption'. Scenario IS92b already assumes that engineers have increased the carbon efficiency of energy use by 1.3% per year, and the ocean and land have adjusted their uptake of carbon. To have net emissions less than 'business as usual', some terms on the right-hand side of Equation (1.3) will have to change. This could be achieved by increasing the amount of low-carbon energy used over that assumed in *business as usual*. Nuclear power is such an example. These alternative energy sources are in general more expensive (when externalities are ignored) than those in use today, and many engineers are sceptical that this is the most cost-effective way to manage atmospheric carbon dioxide concentrations. Land sinks could be changed using ideas such as forestry, a topic treated in Chapter 6, or by creating ocean sinks, a topic treated in Chapter 5. Carbon capture before emission to the atmosphere, and storage, is treated as a decrease in carbon intensity in the discussion below.

Changes in radiative forcing

Radiative forcing of the earth can conveniently be measured in watts per square metre, $W\,m^{-2}$, and compared with the solar irradiance (insolation) at the top of the atmosphere, a value of about 1367 $W\,m^{-2}$. The average solar radiation on a square metre of the earth is much less than this, partially because there is no radiation at night. The changing concentration of carbon dioxide in the atmosphere increases the radiative forcing. The Intergovernmental Panel on Climate Change (IPCC) suggests that the change in radiative forcing of CO_2 due to fossil fuel burning and land changes is about 1.46 $W\,m^{-2}$.

Numerical climate models of the atmosphere–ocean system can be used to predict changes in the earth's surface temperature as a result of changes of radiative forcing. While there is quite a range of values predicted by the present models, a representative result is to predict a change of 1.4–5.8 °C over the next 100 years for greenhouse gas concentrations in the range of Figure 1.2. Remember that, while the radiative forcing changes due to the increase in CO_2, other gases and aerosols may also change the radiative forcing. Aerosols could reflect more solar energy before it reaches the earth's surface. This could, in principle, lead to a net decrease in radiative forcing. In years gone by people worried about a *nuclear winter!*

While we have the difficulties discussed above in predicting the change in greenhouse gas emissions from Equation (1.1), and we have uncertainty in determining the fraction that remains in the atmosphere, we also must use simplified models of the climate system because of the current lack of knowledge of the physics involved. We doubt that our climate models are realistic enough to reliably forecast the future (assuming the emissions and consequential concentration change are correctly predicted), or even whether the change is positive or negative. This is mostly because of the uncertainties of the feedback loop that involves the clouds. Neither can models yet reliably predict under what condition we could expect an abrupt climate change. Such an event, such as a new Ice Age, would be a crisis for humankind. With much of the present arable land under a sheet of ice, a new Ice Age would require major change for many people. Could this adaptation be carried out peacefully?

Managing in the face of uncertainty

How to make the best decision in the face of incomplete information? As the discussion above shows, there is intrinsic uncertainty as to the magnitude of future greenhouse gas emissions and consequential climate change. This is not a matter of obtaining more information about the past. Statistics about the past can tell us about the frequency of events that have happened but cannot be used to predict the future without additional assumptions. Our understanding of social systems and the occurrence of revolutions is totally inadequate for the task at present. In fact, even the examination of possible future emissions and their consequence may lead to a change of human behaviour as we have noted above. The developments in greenhouse gas science and the generation of emission statistics have highlighted the risk of undesirable climate change, which has led to the ratification of the UN Framework Convention on Climate Change by many States. They promise to change the emissions of greenhouse gas away from the *business as usual* model.

Another question we might like to ask is: Why address an uncertain topic like climate change when there are many other pressing problems? This is an issue addressed by Bjørn Lomborg in his book *The Skeptical Environmentalist*, where he suggests that there are global problems with more benefits per dollar spent than climate change.

Consideration should be given to the proposition that the global attention being paid to climate change, rather than some other problem, is the result of random events. When a global champion of some issue captures public attention, this issue is addressed ahead of other more pressing issues that would be identified by a rational socio-economic model. Global inequality, for example, may be a more important issue but it is without an effective champion.

Wigley *et al.* (1996) made reasonable guesses at the land–ocean–atmosphere fluxes of carbon dioxide to convert emissions to concentration that are shown in Figure 1.8. However, they would be the first to agree that issues such as the amount of increase in carbon stored in vegetation due to increased atmospheric concentration of CO_2 are uncertain today. This uncertainty is added to the difficulty of predicting the terms needed to estimate future emissions. Thus there is cumulative uncertainty about the future carbon dioxide concentration that is based on predictions of the level of emission of fossil carbon to the atmosphere.

To say something about the future change in the climate we must include the uncertainty of climate modelling as well. We must rely on the accuracy of a climate model to predict the changes given the predicted carbon dioxide concentration. Here there are many complex processes in the feedback of the climate system that are not well understood. One should not gain comfort from the similarity of predictions from different models. Branches of science adopt paradigms to allow progress. While there is general consensus amongst practitioners of that branch of science, it is the dissenting voice that eventually leads to the overthrow of one paradigm and its replacement with another. Engineers need to be prepared for such shifts, and attempt to incorporate the concept of uncertainty in their solution to humanity's problems.

The role of the engineer is to manage systems in the face of uncertainty. This is done by identifying risks and preparing contingency plans to mitigate the undesirable consequences of events. Based on past experience we can estimate how likely such outcomes might be. We can also estimate the consequences. A very easy strategy for managing the risk that climate change damage is overpredicted is to concentrate on *no regrets* options. These are options that would make sense to take even if greenhouse gas did not create a problem. Changing processes to save energy, for example, can often be a case where the money saved exceeds the cost of the project. The number of these opportunities should not be overestimated. In most well-run businesses these opportunities have already been taken up.

Buying insurance is an approach to managing the risk that climate change will present severe losses. Insurance is the process of distributing the risk amongst a number of parties. There are likely to be some winners and some losers in the impacts of climate change, and a global adaptation fund is of this nature. Developing mitigation technologies in advance of need could also be considered an insurance strategy.

Neither is it necessary to correctly predict the future for an engineer to have acted responsibly. The clarity of hindsight is not available when decisions are made about engineering works. Policy decisions can be 'wise' even if they lead to failure. If the evidence available at the time suggests the decision is the best one, even though it turns out to be disastrous, it is wise. An unreasonable decision, even with an advantageous outcome, is irresponsible. It is failure to have a contingency plan that is not prudent behaviour.

Carbon dioxide emission goals

What are prudent goals for carbon dioxide concentrations? Possibly to stabilise the atmospheric concentration at 550 ppmv. Figure 1.8 shows a prediction of the net emissions needed to achieve a range of concentration in the year 2100. Remember that net emissions are emissions minus the anthropogenic sinks. Stabilising the concentration at 550 ppmv implies a large reduction in net emissions over *business as usual*.

Climate change is presumed to cause economic damage and so economists often talk of stabilising concentration at a level that costs less than the damage caused. Estimating the present value of future damage is controversial.

Many human problems are concerned with food security, and changes in climate are a threat to their food security. Should the concentration of carbon dioxide be managed to minimise disruption to agriculture? This is a different goal to that of optimising global agriculture. Here warming and increased CO_2 fertilisation of crops might lead to a positive change in agricultural output but dramatic losses for some regions. In particular, changing rain patterns may unsettle subsistence farming.

Population, as discussed above, is an element in climate changing emissions of CO_2. Larger population means lower carbon dioxide emissions per person for a particular stabilisation goal. The desire for security of food in old age drives the need for surviving children and the consequently larger families that contribute to overpopulation. Larger families lead to the need for more food, but farm productivity in a rain-reduced area will decline with climate change. To break out of this hopeless spiral of many children for security and reduced food security as a result of climate change due to increased population, the best prospect today is to increase the standard of living and the level of education of the poor. There is a correlation between low standard of living and the number of children in poor families. The interdependence of climate change, poverty and population will be further developed in another monograph.

Once the global net emission goal is agreed, what would be an equitable way to achieve the reductions over those for *business as usual*? Should each person be allowed to release the same amount of carbon dioxide? Should it be based on a country's GDP? Should we use a historical baseline such as 1990? The last position seems to based only on the pragmatic consideration that those with historically large emissions are powerful in world affairs. The second position seems to be based on the argument that the ratio of CO_2 emissions to GDP is a kind of efficiency. Energy is needed to create GDP per person. This suggests that, without an increase in efficiency, the poor in developing counties can maintain low values of emissions only by remaining poor. We will return to this topic in Chapter 7.

What will be the cost of mitigating the net emissions of greenhouse gases to meet the agreed goals? It may be large. Some optimists think that improvements in technology will achieve the reduction at no cost. These technological improvements, when made in the name of climate management are the *no regrets* options mentioned above. As the following chapters will demonstrate, most practical schemes have a significant cost. As well, we should spare a thought for the lost opportunities that occur if resources are diverted to mitigating the emission of carbon dioxide. Unfortunately, a large number of people receive no education, for example, and the resources spent on climate change mitigation could have been used in educating more of these people.

Assessing all the benefits resulting from greenhouse gas mitigation can be complicated. There might be benefits in sustainability that are hard to evaluate. Cleaner air, which might be achieved by moving away from fossil fuel burning, would lead to better health. How can environmental benefit be assessed?

Further reading and exercises

The student can obtain more information by reading books 1, 5 and 7 in the Further reading section at the end of the book.

Exercise 1.1 Take as a scenario that global population will increase at 1% pa, GDPp by 1% pa, and 'greenhouse gas intensity' decrease by 1.3% pa. Graph the predicted emissions of carbon dioxide from 1990 to 2100 (if there were no concern about CO_2). Estimate the amount of carbon dioxide emissions to the atmosphere that needs to be reduced to stabilise atmospheric carbon dioxide concentration. You might like to use Figure 1.8 to estimate the emissions that lead to stabilisation at a chosen concentration level. What is your estimate of the change of land use under the above assumptions?

Exercise 1.2 Water vapour is a strong greenhouse gas. Why is it not considered in the above discussion of emission of greenhouse gas? Should there be more interest in irrigation? Might it be changing the water vapour in the air?

Exercise 1.3 It took about 100 years for the calculations of Arrhenius to lead to proposals for action in the UN Framework Convention for Climate Change. Sketch the history of greenhouse gas understanding, and comment on the possible reasons that action took so long to be an issue.

Exercise 1.4 If we reduced population increase in order to manage greenhouse gas emissions, how much could we afford to spend on each avoided birth? Assume initially that the GDP per person does not depend on population.

Exercise 1.5 Take one of the storylines (scenarios) of the future discussed above and estimate the change in GDPp, population and carbon intensity. Choose one of the emission paths in Figure 1.8 as a goal. How much carbon dioxide needs to be 'avoided' each year if the world were to follow the scenario you choose rather than business as usual? Consider the next 90 years.

2

Changing energy efficiency

Changing carbon dioxide intensity

In the previous chapter we saw how *business as usual* scenarios lead to a rapid increase in the concentration of carbon dioxide in the atmosphere. We saw how GDP per person and population combines with CO_2 intensity to give the total emissions to the atmosphere. The population momentum that comes from past high fertility and the falling mortality makes it difficult for policies to have a short-term impact on population. However, there are some examples such as China, which has changed the rate of increase of its population through regulations known as the one-child policy. Population is a field not usually in the domain where engineers practice. Reducing population growth might be a cost-effective way of managing greenhouse gas, as well as addressing a number of other challenges facing the world, but we will not pursue it further. Readers are referred to Birdsall (1994) for a discussion of population momentum.

The first term, involving the GDP per person, is not one we would advocate reducing to control greenhouse gas emissions. Some assert that the people living in the developed nations consume too much and so have too high a GDP. They are profligate consumers! No-one sensible advocates that the poorest reduce the GDP per person. Would a much fairer distribution of wealth help all to have an acceptable level of GDP with no-one living in poverty? Could the global GDP then be lower than at present and so reduce the emissions of greenhouse gases? These are not questions to which we provide an answer!

Finally the carbon dioxide intensity can be changed. This is a process that has been going on steadily as a result of engineering activity. This chapter looks at the first of a number of ways of reducing the emissions from economic activity at a rate faster than historical experience by increasing the efficiency of energy transfers. We must note that fossil fuel is a finite resource. It will eventually become uneconomic to use as all the easy-to-extract reserves are consumed. Increasing the efficiency of energy transformations reduces the immediate emissions but leaves the fossil fuel for future use. In fact, increased efficiency actually increases the size of the economically recoverable reserves of fuel. Energy efficiency is a way of delaying the increase in atmospheric concentration.

Proposals to increase the energy conversion efficiency to be greater than historical trends are a very popular idea amongst policy makers, but it is not a particularly realistic approach to stabilising carbon dioxide concentrations in the atmosphere. As the models in the previous chapter show, the change in efficiency would need to be faster than the change in population and the growth of GDP per person. This will be hard to achieve. However, efficiency increases will undoubtedly make a contribution to producing lower emission intensity in the future than would have occurred without concern about increasing CO_2 concentrations. Changing CO_2 intensity is the position announced by the President of the USA in 2002 as the USA response to the rising levels of carbon dioxide. Remember we assumed in *business as usual* that the carbon dioxide emitted in producing the GNP would decrease by 1.3% per year for the next 100 years. A greater reduction than this is required to lower the carbon dioxide concentrations from that expected in the *business as usual* scenario.

This is a good time to note that changing anthropogenic emissions is not the only strategy for managing carbon dioxide concentration in the atmosphere. One can also change the anthropogenic sinks of carbon dioxide. A number of factors can influence the net flux of carbon dioxide into the air, and we will treat them in the following chapters. In this chapter we look at how the energy intensity might decrease by generating energy from fossil carbon more efficiently and putting it to work more effectively. It is not energy that people want; it is the result of the application of it, e.g. warmth or aluminium, that is desired. To have efficiency rise more rapidly than 1.3% pa will require the introduction of new technology at a faster rate than is expected from historical trends. Can governments design policies to achieve this and can engineers provide the innovations needed?

Engineers can almost always increase efficiency of energy usage by increased capital expenditure. For new plants, the economically attractive level of efficiency may be higher than present practice, but, for existing plants, changing the efficiency may represent a financial problem. Energy conversion plants represent a capital investment with a lifetime of 30 years or so. New plants are seldom so much more efficient than their predecessors that scrapping the existing capital investment can be justified on economic grounds. Thus change is slow and there will always be a tail of less efficient plants. Attractive innovations are those that produce greater efficiency for a cost that justifies the investment. Such a solution is termed *no regrets*.

In a regime where there is a cost for dumping carbon dioxide in the atmosphere, it will be possible to design and commission processes that operate at higher energy efficiency and more (capital) cost. In this case the saving from the reduced dumping cost can justify a higher capital cost than the optimum process with free carbon dioxide dumping. Thus we expect that plants in the USA designed in 2008 will be a little less efficient than if there were a charge for carbon dioxide emissions to the atmosphere. Before we can go much further we need to be more careful about our use of the word efficiency.

Energy is one of the basic commodities of a modern society and is often managed by the government rather than left to *free market* forces. This is true in both centrally managed economies such as China and in *free market* countries such as the USA. The result of restricting the cost of electricity to consumers is political popularity, but often leads to supply shortages. Low prices inhibit capital investment in new plants, which we expect, from long experience, to be more efficient than existing plants. Regulation of energy can produce *distortions* in the market that lead to more greenhouse gas emissions.

Next is the question of the cost of providing primary energy. The common approach of considering only the *direct* costs of a project is a problem for engineers who are trying to provide the 'best' solution to society. In many cases in the past, the cost to the environment of projects was treated as an externality. This means it was someone else's problem. Thus the costs in terms of climate change as a result of dumping carbon dioxide in the atmosphere have not in the past been considered in estimating the level of capital expenditure (and efficiency) for the best financial return.

As the primary energy used to fuel economies shifted from human exertion, to animal power and then to fossil fuel, there has been an increase in the anthropogenic emissions of carbon dioxide. Fossil fuels are rich in carbon, and combustion releases heat (energy) in the reaction

$$C + O_2 = CO_2 + heat$$

Since the molecular weight of C is 12 and of O is 16, it is easy to see that 12 tonnes of carbon produces 44 t of CO_2. A typical value of carbon dioxide emission from electrical energy generated from coal is 900 $gCO_2\ kWh^{-1}$.

While coal is mostly carbon, natural gas, for example, is mostly methane, CH_4, which generates heat, in addition to the oxidation of the carbon, by converting the hydrogen to water. Thus, for the same heat, methane produces less carbon dioxide: typically 375 $gCO_2\ kWh^{-1}$. The use of natural gas does not necessarily lead to a more efficient transfer of energy; rather the energy contained in the hydrogen does not produce carbon dioxide. Substituting hydrogen-rich fossil fuels is the basis of fuel switching strategies that reduce the carbon dioxide intensity per unit of GDP.

Much attention has been focused on the concentration of carbon dioxide in the flue gases of power stations and the possible re-use of this chemical compound (e.g. Riemer, 1996). The carbon dioxide in current combustion practice represents some 15% of the exhaust gases. Smaller amounts are captured in industrial processes such as ammonia manufacture and are available in concentrated form. Since carbon dioxide has given up its energy, it is of no value as an energy source and is in low demand as an industrial product. Steinberg (1999) has looked at recycling some of the carbon dioxide as carbon black, but for this use to be significant it requires the establishment of large new markets.

Sectorial distribution of greenhouse gas emissions

Which sectors of the economy lead to large emissions of greenhouse gas? We will use the year 1990 as a reference in line with the UN Framework Convention on Climate Change. Combustion used about 6000 Mt yr^{-1} of carbon which produced about $6 \times 44/12 = 22$ Gt of CO_2. Land use changes were more uncertain. Estimates place emissions at 1.1 to 3.6 GtC yr^{-1}. Notice it is common practice to express the amount of carbon dioxide either as tonnes of carbon dioxide or as tonnes of carbon contained by the carbon dioxide. Prefixes such as mega = 10^6 are listed in Appendix 5.

The Intergovernmental Panel on Climate Change (IPCC), in its *Summary for Policymakers – Mitigation* have made an estimate of the potential greenhouse gas net emissions from different economic sectors in the 2010 to 2020 time frame. We have reproduced some of their findings in Table 2.1. As well, the table contains estimates of the potential emission reductions that could be achieved. Half of these could be made as *no regrets* options, with buildings providing a significant opportunity to be modified or designed to consume less energy. Transport emissions, although not so large a source of carbon dioxide at present, are rising rapidly.

Note that the total includes emissions from other sectors, notably agriculture, which are not included in the table. This table suggests yearly emission reductions of the order of 12 $GtCO_2$ are possible in 2020 at *manageable* cost. This is keeping emissions more or less constant at 1990 values.

The volume of concrete produced is second only to the water used by humankind. The manufacture of cement needed for concrete is a big producer of carbon dioxide. The process involves separating calcium carbonate, of molecular weight 100, by the reaction $CaCO_3 \rightarrow CaO + CO_2$, into calcium oxide and carbon dioxide, molecular weight 44. This process produces about $44/100 = 0.44$ tonnes of CO_2 per tonne of limestone. The term for calcium oxide which contains other impurities (about 30%) is clinker. Clinker is unground cement. Thus the 70% CaO in cement releases about 0.44×0.7 tonnes of CO_2. Fuel is needed to produce the above reaction, and many

Table 2.1. *Estimates in the year 2000 of potential global greenhouse gas emission reductions in 2010 and in the year 2020.*

Sector	Historic emissions in 1990 (MtC_{eq} yr^{-1})	Historic C_{eq} annual growth rate in 1990–1995 (%)	Potential emission reduction in 2010 (MtC_{eq} yr^{-1})	Potential emission reduction in 2020 (MtC_{eq} yr^{-1})	Net direct costs per tonne of carbon avoided
Buildings CO_2 only	1650	1.0	700–750	1000–1100	Most reductions are available at negative net direct costs
Transport CO_2 only	1080	2.4	100–300	300–700	Most studies indicate net direct costs less than \$25 tC^{-1}, but two suggest net direct costs will exceed \$50 tC^{-1}
Industry CO_2 only	2300	0.4			
– energy efficiency			300–500	700–900	More than half available at net negative direct costs
– material efficiency			~200	~600	Costs are uncertain
CO_2 total	5200				
Industry Non-CO_2 gases	170		~100	~100	N_2O emissions reduction costs are \$0–\$10 tC_{eq}^{-1}
Total	6900–8400				

MtC_{eq} = megatonnes of carbon equivalent.
Source: Metz *et al.* (2001).

fuels can be used. A typical figure is 5.9 GJ t^{-1} of clinker for wet kilns, and 3.5 GJ t^{-1} for dry kilns. As well as fuel, some electricity is used in operating the kiln. A typical figure is 0.3 GJ t^{-1}. (Remember 1 GJ = 278 kWh.) A simple calculation gives an order of magnitude estimate of CO_2 emissions (Table 2.2).

Cement manufacture is rising steadily, and in the year 2002 exceeded 1800 million tonnes. This suggests emissions from manufacture of order 1.5 Gt yr^{-1} of CO_2. Some of this carbon dioxide is reabsorbed over time when the cement is used in concrete. There are opportunities for reducing emissions by substituting low-carbon fossil fuels and capturing the carbon dioxide from the limestone in the kiln flue.

Table 2.2. *Carbon dioxide emitted per tonne of cement.*

			Carbon dioxide generated (t)
CO_2 emission from $CaCO_3$ conversion	0.44×0.7 t		0.31
Heat from coal	3.5 GJ		0.39
Electricity	0.3 GJ	83 kWh	0.08
Total			0.78

typical coal = 0.11 kg CO_2 per MJ (lower heating value).

Another industrial process important in the supply of food is the manufacture of reactive nitrogen in the form of ammonia as a fertiliser for crops. Forty per cent of the protein consumed by humans today relies on the Haber–Bosch process to synthesise ammonia from fossil fuel energy sources. This dependence on nitrogen is part of the *Green Revolution*. The energy efficiency of the production of ammonia is increasing, and a modern ammonia plant uses about 28 GJ of natural gas to produce one tonne of ammonia. Older plants and those using other fossil fuels such as coal are less energy efficient. However, the rising price of natural gas is changing the relative costs of production and so-called clean coal is likely to be more important in the future. China already has a number of ammonia plants running on gasified coal. Worldwide about 100 million tonnes of ammonia are produced. Let us assume an average energy demand of 30 GJ t^{-1} and deduce that ammonia manufacture uses about 3×10^9 GJ of energy per year. Appendix 5 provides typical values of the energy of natural gas. If it is natural gas producing the energy, one GJ is equivalent to 18 kg of gas. We can see that methane, CH_4, with a molecular weight of 16, produces 44/16 = 2.75 times its weight in CO_2. Thus the annual carbon dioxide produced is about 150×10^6 tCO_2. The total emissions from fossil fuels in the year 2000 were of the order of 25 $GtCO_2$. Ammonia manufacture accounts for some 0.6% of emissions. Not all of this carbon dioxide is dumped in the atmosphere, as some is used in the manufacture of urea from ammonia. Urea is a popular agricultural fertiliser, and most of the CO_2 is released when the urea is distributed on the soil.

Swaminathan and Sukalac (2004) suggest that if best available practice, BAP, were employed, the carbon dioxide emissions from reactive nitrogen manufacture could drop to 42% of their current value.

An emerging new energy demand is for water desalination. The rising world population is increasing demand for fresh water, while the supply remains fixed. Much of the conveniently located water is already fully exploited. In developed countries, the environmental value of flowing rivers and a hostility to large dams restricts the new supplies of fresh water.

A modern desalination plant using membrane technology has recently been commissioned at Ashkelon, Israel. It can produce $100 \times 10^6 m^3 yr^{-1}$ of fresh water, some 5% of Israel's total water usage. The capital cost was of the order US$250 million, and cost of operation, maintenance and capital serving is about US$0.53 per cubic metre. The plant occupies 75 000 m^2 of industrial land.

Electrical energy required is stated to be 4 kWh per cubic metre of water. This is about 4000×3600 joules per cubic metre and would require about 50 MW of generating capacity. Assume the gas turbine figures of page 27 (375×10^{-6} tCO_2 kWh^{-1}); the total CO_2 generated by this operation is $4 \times 375 \times 10^{-6} \times 100 \times 10^6$ tonnes of CO_2 per year (0.15 million tonnes per year). Desalination of seawater is opposed by some when the carbon dioxide produced is disposed of in the atmosphere.

There are many other industrial processes that use large amounts of energy; the production of steel and other metals are examples that continue to grow. In 2006 this sector produced about 1 300 million tonnes of steel and had an average emission rate of 1.7 tCO_2 per tonne of steel. This figure could be greatly reduced by using new technology.

Since heating, lighting and cooling of buildings are significant sources of carbon dioxide, two case studies of building efficiency are included later in the chapter. Energy use by buildings continues to increase despite more energy-efficient designs, in part because the occupiers of buildings demand more space and more appliances. There remains, however, the potential to reduce carbon emissions by about 4 $GtCO_2$ yr^{-1} over *business as usual*, according to the IPCC.

Heat loss from buildings

Much of the greenhouse gas from buildings comes from the energy needed to heat or cool offices and houses. The heat load on buildings depends in part on the solar energy entering through the windows. The glass can have a coating that regulates the solar energy and visible light entering. Small solar cells provide the control. Smart windows may be more efficient than solar cells (a so-called renewable energy source with low CO_2 emissions discussed in the next chapter) as a way of managing the internal temperature of a building.

A simple way to reduce the energy used in building heating is to adjust the temperature, accepting a hotter temperature in summer and a lower temperature in winter. This approach will be most effective in developed countries. The costs can be low (intelligent thermostats) and so this can be a *no regrets* option. The issue is convincing people that there is no loss in wellbeing.

When energy generation is carried out with increased efficiency (as discussed in Chapter 3), the greenhouse gas load on the atmosphere decreases. As we reduce greenhouse gas from energy generation (that is, its CO_2 intensity decreases) the greenhouse gas saved by the reduction of energy from buildings becomes less. If, as we discuss below, primary energy were to be produced with (near) zero emissions, there would be no reduction of greenhouse gas through changes in energy consumed in buildings.

Another source of greenhouse gas in the building sector is the waste from cooking stoves in developing countries. Remember, firewood is not a contributor to the build-up of carbon dioxide in the atmosphere but petroleum is.

Efficiency of energy transformation

When heat is converted to work such as by a boiler and steam turbine, there is less work done than the heat supplied. The heat starts at some temperature T_1, and as it does work it gives up heat to reach temperature T_2. The most efficient conversion is the Carnot cycle, and it has a theoretical efficiency of

$$\eta = 1 - \frac{T_2}{T_1} \tag{2.1}$$

It can be seen that, the higher the initial temperature, the greater the efficiency, and so this has driven the ever increasing temperature used in steam turbines to produce electricity. The development of high-temperature materials for turbines paces the increase in inlet temperature. Typical initial temperatures today are of 810 K, and exit temperatures of 293 K.

Electricity is a very convenient way to deliver energy through countrywide high-voltage transmission networks. Electricity is a high-value form of energy. As can be seen, the transformation of heat to electricity wastes a considerable amount of energy. Would the conversion to another energy carrier such as hydrogen be better?

The change in carbon dioxide generated by burning carbon more efficiently to generate electricity can be estimated as follows. Assume the initial energy conversion efficiency is η. Then carbon that contained H units of heat produces ηH units of electricity. Now, if the efficiency becomes $\eta + \Delta\eta$, the electricity becomes $(\eta + \Delta\eta)H$. As the amount of carbon burned is the same, the carbon dioxide generated remains the same. If one only wants ηH units of electricity, the carbon dioxide generated in the case of the higher efficiency unit is the fraction $\eta H/(\eta + \Delta\eta)H = \eta/(\eta + \Delta\eta)$ of that generated by the lower efficiency unit. If $\eta = 0.4$, then a 1% increase in efficiency reduces the carbon dioxide generated to 0.4/(0.4 + 0.01), a reduction of 2.5%.

Cogeneration is the term used for the combination of electricity generation and the supply of low-temperature heat for uses such as domestic heating or industrial drying. The energy contained in steam or gas at T_2 from the power station need not

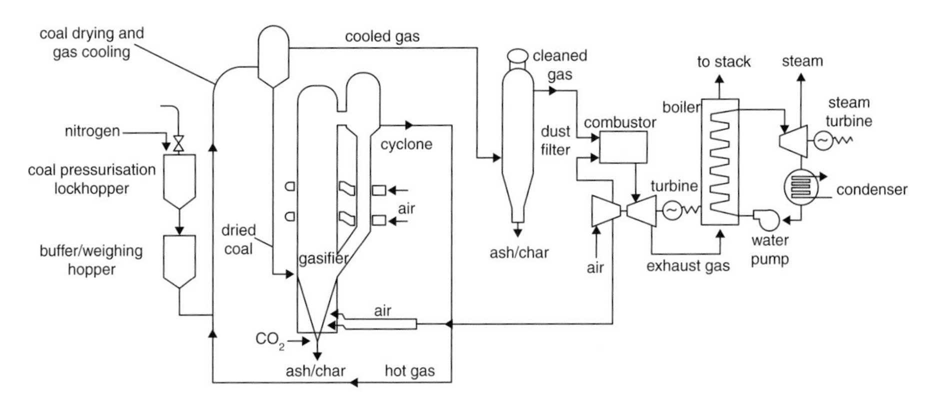

Figure 2.1 A coal gasification electric power generation system known as IGCC (integrated gasification combined cycle) after Smith, 1999.

be all wasted. The use of the low-temperature energy needs to be near the power station, as the losses in moving low-temperature heat are in general more than moving electricity through a network of wires. The efficiency in using fossil fuel can be increased substantially, as a modern boiler converts fossil fuel to heat with about 90% efficiency. There are opportunities of 20 to 30% reductions in carbon dioxide emissions by cogeneration.

Since there are large reserves of coal there have been research programmes under the banner of clean coal to increase the efficiency of converting coal to useable energy. These efforts have mostly focused on gasification. The concept is illustrated in Figure 2.1.

A still newer concept is underground coal gasification which replaces the coal-mine and the surface gasifier stages, which are both large capital items. This offers the prospect of producing syngas more cheaply than at present.

Once the fossil fuel is converted to work, the efficiency with which that work is utilised is important. When fossil fuel is used directly for heating, energy escapes from where it is needed. For example, a traditional fireplace creates a draught of air which carries much of the heat up the chimney and is wasted for heating the people. Insulation influences the efficiency with which fossil fuel burning provides the required heating. If energy is used for cooling, again the house or factory insulation is important.

Thus there are two main components to the efficiency of energy usage:

that in converting some source of energy such as natural gas to electricity or work or heat; and

that of converting the electricity or heat to some commodity such as refrigeration or lift operation or light.

Both components offer potential to increase efficiency and, when the primary source of energy is fossil fuel, to reduce greenhouse gas emissions.

There is also much scope for increased efficiency of home appliances. Low capital cost devices are often energy inefficient. The most common approach to this issue is to require manufacturers to show the energy efficiency so the consumer can make an informed choice. Standby power in devices such as TV or computers may be substantial and also offers scope for increased efficiency.

Efficiency of transport of energy

Coal became an important source of energy in the Industrial Revolution because of its low transportation costs, but costs are lower for uranium, the energy source for nuclear power reactors. Representative costs are:

Uranium at order 600 GJ kg^{-1}; transport cost is low.
Coal at 33 GJ t^{-1}; transport:
by sea: US$0.06 per GJ per 1000 km (in 2004; this price is quite volatile);
by train: US$0.25 per GJ per 1000 km.
Natural gas by pipeline: US$1.0 per GJ per 1000 km.
Electricity direct cost: US$1.0 per GJ over 1000 km (representative electricity cost = US$0.05 per kWh = US$14 per GJ).
(1 kWh = 1000 × 3600 J; 1 GJ = 278 kWh.)

Case study: retrofitting power stations

Many electric power stations use steam turbines that operate at low steam pressure and temperature. This is known as a subcritical boiler plant. When the boiler operates above the saturation temperature of water vapour, the vapour is known as superheated steam. There is a critical pressure at which the liquid phase (water) has the same density as the steam (vapour) and it changes straight into vapour. Boilers operating above the critical point are known as supercritical boilers (although there is no actual boiling in this case).

If we 'repowered' existing low-temperature power stations with supercritical boilers, the efficiency of the electric power station would be increased. In Equation (2.1), T_2/T_1 would be smaller. All the items needed for the upgrade are available commercially. Typically the efficiency of subcritical boilers is 34% based on the lower heating value of the coal. The heating value of the fuel when the water in the combustion gases is a vapour is the lower heating value (LHV). (Typical LHV of carbon is 33 GJ t^{-1}.) Since a more efficient boiler uses less coal to produce the same amount of electricity, the emissions of carbon dioxide per unit of electricity are reduced. Loh *et al.* (1998) have examined the economics of making such refits in the USA.

They took as a reference a 250 MW power plant having a drum-type natural circulation pulverised coal boiler. It was assumed that it operated at a pressure of 165 bar (1 bar = 1 atmosphere) and a temperature of 540 °C. They considered an alternative of a supercritical boiler operating at 310 bar pressure (31.4 MPa) and 594 °C. Let use assume the condenser temperature, T_2 is 280 °C. In this case the Carnot efficiency for the original boiler would be 48%, while the new system has a Carnot efficiency of 53%. Remember that the Carnot efficiency is not achieved in practice.

The same steam turbine is used, but a new boiler and an additional new peaking turbine, which is shown in Figure 2.2, are employed. Loh *et al.* (1998) took the original plant efficiency as 38.0% (LHV) and 39.7% (LHV) for the new system; the change is 1.7%. Efficiencies of car engines and steam turbines are normally based on lower heating values, since water normally leaves as a vapour in the exhaust gases, and it is not practical to try to recover the heat of vaporisation. As a rule of thumb taken from Figure 2.3, an improvement in efficiency of 1% gives a 2.5% reduction in CO_2. The same amount of electricity can be generated with less coal. Thus we expect CO_2 emissions to be reduced by $2.5 \times 1.7\% = 4.25\%$. Many representative values are presented in Smith (1999).

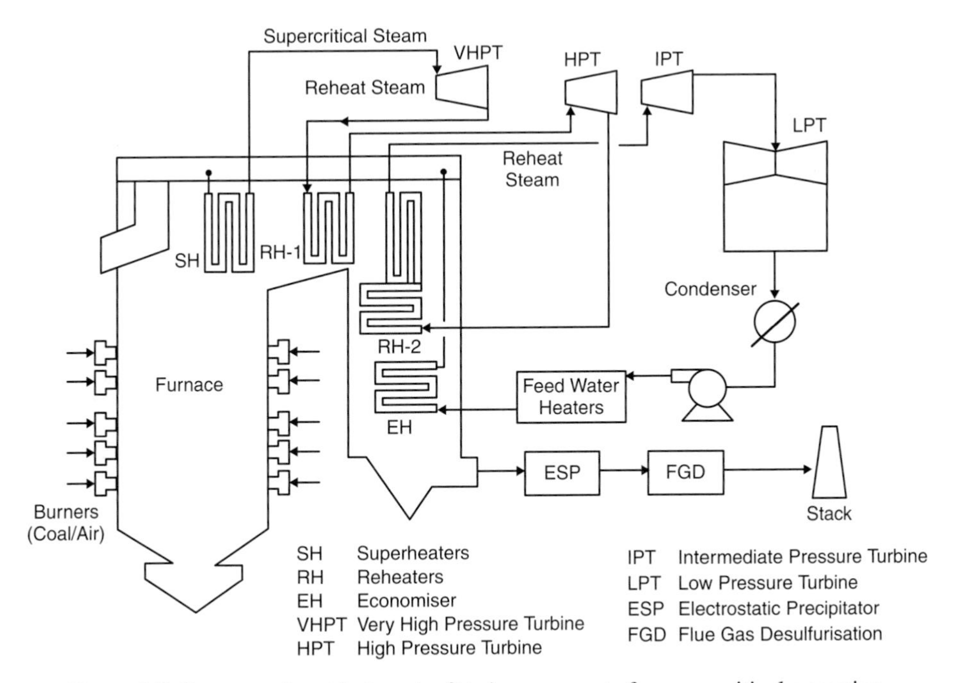

Figure 2.2 Steam cycle with the retrofitted components for supercritical operation.

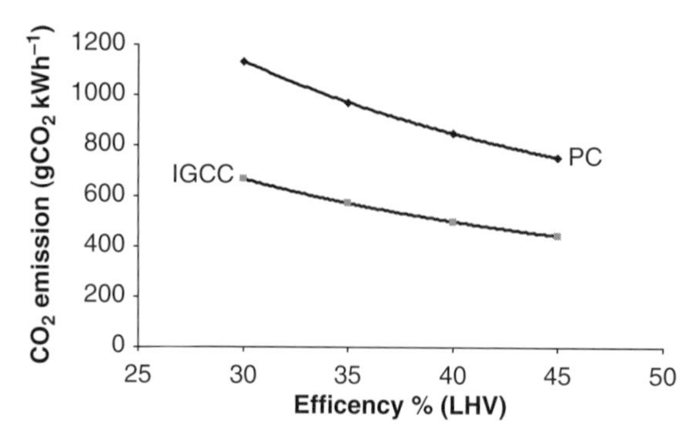

Figure 2.3 CO_2 release as a function of efficiency for different power generation systems. IGCC = integrated gasification combined cycle; PC = pulverised coal.

The cost of electricity depends upon the cost of the capital (interest) and the expenses of running the power station. The latter are mostly the cost of the coal. For typical heating values of coal, and efficiencies of about 40%, one needs about 0.25 kg of coal per kWh of electricity.

The cost of such an upgrade is estimated at US$180 per kW. We need to estimate the interest on such an expenditure. Say we take 10% pa to cover interest and depreciation. This is a yearly cost of US$18 per kW. With the increased efficiency, the plant will use less coal per unit of power produced. Also the operational costs such as supervision and maintenance will not change much so we will assume they also decrease per unit of power. Thus, increased efficiency leads to a decrease of the operating cost per unit of power by about $1.7/38 = 0.045$ of its earlier value. If we estimate the cost of the coal at US$0.01 per kWh (US$40 per tonne), we have a yearly coal cost of US$65.7 per kW. The increased efficiency saves US$2.96 per kW of this cost. Not enough to justify the capital expenditure, which requires $18 per year to service the capital. This is why most subcritical boiler operators have not increased their efficiency.

Let us assume that the plant operates at 75% capacity, producing for each kW of capacity $24 \times 365 \times 0.75 = 6570$ kWh of energy per year. It is also known that the production of 1 kWh produces about 1.0 kg of CO_2. Over a year this is $1.0 \times 6570 = 6570$ kg of CO_2. A 1.7% increase in efficiency decreases this by $0.0425 \times 6570 = 279$ kg per year. Thus we can calculate the cost of making this reduction in carbon dioxide. It is US$18, less the operating saving of US$2.96 = US$15.04. Thus a reduction of 279 kg CO_2 costs US$15.05. One tonne of CO_2 reduction costs US$54 per tonne of CO_2. As we will see later, this is a relatively expensive way to reduce carbon dioxide. However, it is possible today and has almost no undesirable side effects. The same land and infrastructure are used.

This example makes the point that engineers can increase efficiency but it is not an attractive economic action (assuming CO_2 can be exhausted to the atmosphere without cost). Most large-scale organisations are aware of the relative costs of savings in material or labour and are operating near the optimum efficiency, as they see it. This is not true of smaller operations that do not have a strong involvement of engineers. Residential building is such a situation.

Case study: building heating and air conditioning

This is a situation where minimisation of capital costs has led to poor efficiencies. Air conditioners of low efficiency are cheaper, in general, to purchase, especially those with a heating cycle. If decisions about domestic air conditioning are made by people able to consider rational economics, there is potentially scope to reduce greenhouse gases as a *no regrets* option.

Heat pump systems provide two- or three-times as much heat as the electrical energy they use to move the heat. They work by compressing and then expanding a gas in a closed cycle. Usually such systems, when heating the building, expand gas outside the house and transfer heat from the outside environment to the gas before compressing the gas and transferring the heat, now at a higher temperature, to the area inside the house. They are often known as reverse cycle units. Let us consider the case of a house that uses a 1 kW strip heater or electrical wall heater (radiator). If it ran continuously for 3 months per year, say 100 days per year, the electrical consumption would be 2400 kWh per year. At a domestic cost of electricity of order US$0.1 per kWh this is US$240 per year.

To compare the two choices, a strip radiant heater powered by electricity and a heat pump, one needs to consider the capital cost, the expenses incurred and the time value of money. We will work in constant dollars so that inflation is not an issue. The time value of money recognises that an expense in the future is less painful than an expense now. Let us assume we borrow the capital at 10% interest and assume the allowance for depreciation is kept in an interest free account to replace the unit at the end of its life. Some of the details are set out in Table 2.3

It can be seen that in this example the heat pump saves 1400 kg per year of carbon dioxide emissions. It also saves the householder US$66.5 per year. This is a *no regrets* option. Over the lifetime of the heat pump it would save 7000 kg of CO_2. We do not need to consider the time value of money in this simple example.

Say the government decided it would charge US$20 per tonne of CO_2 for the emissions. This is US$0.02 per 1.0 kg of CO_2, and we know from our rule of thumb for electricity generation from coal that 1.0 kg of CO_2 comes from 1 kWh of energy. If electricity consumers had to pay this, it would represent about a 20% increase from (a typical) US$0.1 per kWh. Alternatively it might fund the emission reduction from taxation. The other option is for the government to provide

Table 2.3. *Heat pump benefits (1 kW capacity) in US$.*

	Strip heater	Heat pump
Capital cost	$10	$250
Installation cost	Neglected	Neglected
Yearly operating cost (100 days)	$240	$100
Efficiency[a]	100%	240%
Life	20 years	5 years
Interest pa	$1	$25
Linear depreciation	$0.5 pa	$50 pa
Yearly cost	$241.5	$175
CO_2 emissions (1 kg kWh^{-1})	2400 kg	1000 kg

[a] ratio of heat to electrical energy.

Table 2.4. *Discounted emission reductions.*

Year	Carbon saved (kg)	Discounted value of carbon (kg)
1	1400	1400
2	1400	1330
3	1400	1260
4	1400	1200
5	1400	1140
Sum	7000	6330

For more information about present net value see Appendix 2.

a subsidy to buy the heat pump. The avoided carbon dioxide is worth US$28 per year in this example, but we note that the carbon dioxide saving is delivered over a number of years. If the subsidy were made available on purchase of the heat pump, it could discount the value of the carbon dioxide in future years. Let us discount the first year's CO_2 saving by 10% and subsequent years by a further 10%. The subsidy could be $28 \times 0.9 + 28 \times 0.81 + 28 \times 0.73 + 28 \times 0.66 + 28 \times 0.59 =$ US$103.32. Such a subsidy would be an additional incentive to the householder. However the householder may still not want to go into debt or invest the US$200 in the heat pump. Why is this?

Another way to consider the problem is to calculate the equivalent 'present amount' of carbon dioxide avoided. Since not all the savings in CO_2 emissions come in year 1, we would like to be able to compare options with savings spread out over different time periods. Take a discount rate of 5% pa and no discount in the first year. (The rate of discount is controversial.) Then consider Table 2.4.

Thus the present value of five years of carbon savings is 6330 kg. A US$240 subsidy on purchase of the heat pump (its increased cost over the alternative) would amount to about 240 / 6.33 = US$38 per tonne of carbon dioxide.

Case study: lighting

Electric lighting is used extensively throughout the world. The incandescent bulb works by heating a thin tungsten wire in a glass bulb filled with inert gas. The wire is 'white hot' and radiates much energy outside the visible spectrum. This heat is wasted if the desire is to produce light. In hot climates and wealthy countries, air conditioning is used in part to remove this heat.

An incandescent bulb that uses 60 watts of electricity produces about 550 lumens of light. A compact fluorescent bulb, however, can produce the same light with only 11 watts of electricity. Thus there is a saving of 49 watts per hour of operation. Expressed another way, after 20 hours of operation the new light bulb will have saved 1 kWh of electricity or about 1 kg of CO_2. What is the cost of making this reduction in greenhouse gas?

Compact fluorescent bulbs have both a longer lifetime and use less electricity. It could be that the long-life fluorescent globes make better economic sense than cheaper incandescent bulbs. If this were so, the saving in CO_2 emissions would have been made not only at no cost but with an economic profit. This, we are getting to know, is a *no regrets* option to reduce carbon dioxide. It means that one is pleased to make this change even if avoiding increasing carbon dioxide concentration in the atmosphere is not necessary. You will have *no regrets* that you changed to compact fluorescents because you are saving money.

Why is this change from incandescent bulbs not complete? One reason is that consumers are not rational economic entities. The fluorescent bulb costs more than an incandescent bulb to buy. Also the economic advantages are not self-evident. You need to believe that the fluorescent bulb will outlast a number of the cheaper incandescent bulbs.

CO_2 intensity as a function of GDP

One can understand that increasing the efficiency of an energy-intensive process has the potential to lower the CO_2 emitted for each unit of GDP. The present situation can be examined in Figure 2.4, which is drawn from Table 2.5. Here we use gross national income rather than GDP.

One sees from Table 2.5 that a developing country such as India uses energy rather inefficiently to generate GDP. Some care is needed, as we noted earlier, since

Table 2.5. *Carbon dioxide emissions from selected countries.*

Country	GNI US$billion US (2002)	CO_2 MtC yr^{-1} (2002)	Intensity kgC yr^{-1} US $\$^{-1}$
Australia	386	86	0.22
India	501	334	0.67
Japan	4 265	330	0.08
Germany	1 870	218	0.12
USA	10 110	1627	0.16

CO_2 figures from http://cdiac.esd.ornl.gov.
GNI figures from www.worldbank.org/data/databytopic/GNI.pdf.

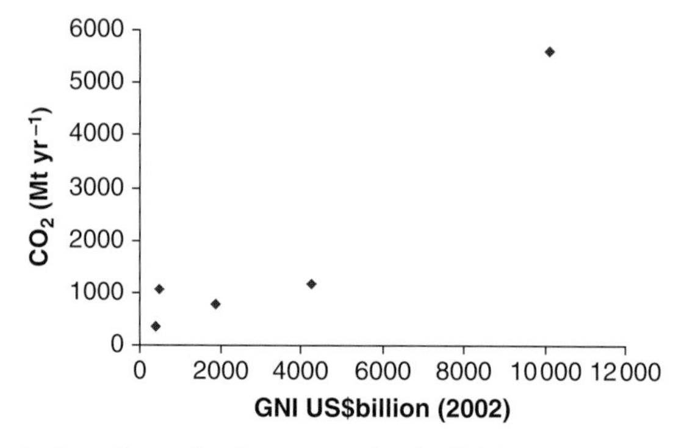

Figure 2.4 Emissions from the five countries in Table 2.5 as a function of gross national income.

GDP does not capture subsistence activities at all well. Some developed countries, like Japan, are much more frugal with energy than say Australia or the USA.

The gross carbon dioxide intensity can be changed by three approaches: changes in energy efficiency (e.g. use less energy for the same result); changes in the carbon dioxide emission from the generation of primary energy (e.g. fuel switching, discussed in Chapter 3, or carbon capture and geological storage); or changes in lifestyle so as to not consume as much energy (e.g. bicycle riding). Efforts to change the lifestyle are not usually an issue confronted by engineers. The lifestyle changes can be brought about by economic persuasion or by moral persuasion. The first can be thought of as a carbon tax. Here the government makes it more expensive for the consumer to purchase carbon-based energy, and so demand is expected to fall. Moral persuasion is exhortation to save the environment, and large international meetings such as the Rio meeting in 1992 had the aim of changing people's view of the use of finite resources such as coal and the value of the 'natural' environment. The Rio meeting we suspect had little impact on the poor trying to live on

US$1 per day. It also has little impact on the affluent who are acting to optimise their economic position. (Do you think the efforts to encourage voluntary activity to reduce greenhouse gas emissions are cynically proposed because it is suspected they will be ineffective?)

Much of the increase in standard of living has been achieved by the consumption of energy. It involved substituting energy for labour. Where men used to dig trenches with shovels they now use petroleum-fuelled backhoes to dig a much longer trench in a shorter time. Now the question arises as to the alternatives uses of the free time generated. It may be to work for additional income, which in general produces greenhouse gas, or it may be to undertake some unpaid but attractive activity such as to walk in the park. The latter produces no greenhouse gas. These are lifestyle choices. As each individual experiences a different opportunity cost, it is hard to know the savings in the cost in energy that is required to motivate people to, say, spend an hour changing their light bulbs to compact fluorescent bulbs. Moral persuasion can be used to convince you to undertake such activity even if it is not as economically rewarding as, for instance, working for an hour where you receive income.

Further reading and exercises

The student can learn more from item 11 of the Further reading section.

Exercise 2.1 Consider a black coal electric power station. Its present efficiency is 35% of the LHV. The lower heating value of coal is about 33 000 kJ kg^{-1}. What is the implication for greenhouse gas emissions for an increase in efficiency? Increasing efficiency will require new capital expenditure. Assume each 1% increase in efficiency costs 5% of the existing capital. Assume that 40% of the cost of power generation is coal, 40% is the cost of capital and 20% is fixed (for maintenance etc.); what is the cost of abating carbon dioxide by changing efficiency? (Useful rule of thumb: 1 tonne of coal = 1.5 tonnes of CO_2. Can you suggest why it is not in the ratio of molecular weights, 44/12?)

3

Zero-emission technologies

In the previous chapter we looked at the concept of increasing the efficiency with which fossil fuels were used to produce work. However, this increase in efficiency will need to be taken up at an unprecedented rate under most of the scenarios discussed in Chapter 1 in order to stabilise greenhouse gas concentrations in the atmosphere. As an alternative, we could deploy technologies that emit near-zero greenhouse gas. This would also have the effect of reducing the carbon dioxide intensity.

About 20% of the world's primary energy at present comes from sources that emit no **net** carbon dioxide. Firewood is the largest component. Here carbon dioxide is extracted from the atmosphere by photosynthesis as the tree grows and is returned to the atmosphere during combustion. The primary source of energy for firewood is the sun, which radiates energy to the earth. The next most important low-emissions energy sources are nuclear energy and hydro-energy, which are about equal contributors to energy supply.

We need to recall the magnitude of the reduction in net carbon dioxide emissions required to stabilise the atmospheric concentration at some value. The emissions are shown in Figure 1.8. For example, if we assume the scenario IS92a is a *business as usual* scenario for the world without concern for greenhouse gas, then to achieve stabilisation of concentration at 550 ppm after 100 years as per Figure 1.8, the world **net** emissions need to be held to about 9 $GtC\,yr^{-1}$ in 2030 and have a lower value thereafter. This is a reduction in **net** emissions by 2 $GtC\,yr^{-1}$ in 2030, which is less than half the additional energy capacity projected to be required to achieve sustained economic growth. Net emissions are anthropogenic emissions minus anthropogenic sinks (as defined in Chapter 1, Equation (1.3)). How much of this 8 $GtCO_2\,yr^{-1}$ can be achieved by introducing more primary energy sources from which little carbon dioxide is emitted to the atmosphere?

Nuclear power

Nuclear energy is a zero-emission technology (although the mining of the fuel is not) which is currently cost competitive with fossil-fuel power plants. Here we have assumed that the costing has captured all the externalities. We will return to this point.

There is a very large known supply of nuclear fuel (uranium), but, like most resources, the amount that is economically attractive to extract depends on the price offered. There appear to be enough reserves (known resources whose magnitude has been assessed), so burning of uranium is a sustainable activity for the next 30 years. However, nuclear power is not a popular option with some sections of the public. Here we mean some sections of the one billion or so people currently living a comfortable life. Those living in poverty presumably do not have an opinion. Concerns focus on risk of an accident such as occurred at Chernobyl in the Ukraine. The World Health Organization (2006) report that, as well as the immediate deaths, thyroid cancer is much more prevalent in the area subjected to radiation from the accident. An additional concern is the ability to use nuclear power generation as a cover for constructing nuclear bombs. Since the technologies used in nuclear power generation overlap with those used in nuclear weapons, there is concern about 'proliferation'. The reprocessing of spent fuel is another controversial point.

The longer planning times for nuclear power plants, due to the public interest as well as the high capital costs, may discourage investors who prefer investments with shorter payback times. The initial capital costs may be very high, but the fuel costs are lower than for fossil-fuel power stations. This means that the total generating costs are less susceptible to fuel price fluctuations. As nuclear power is a mature technology, a number of countries, notably France and Japan, are planning to increase their nuclear capacity. Present installed capacity is about 350 GW ($= 2.6 \times 10^{15}$ kWh per year, see Appendix 5 for conversion factors). If this amount of electrical energy were provided by coal, CO_2 emissions at 1 kg kWh^{-1} would be 2.6 GtCO_2 yr^{-1}. Increasing the nuclear power by a factor of three by 2030 would achieve our emission reduction goal above.

The mining of the uranium ore is a generator of carbon dioxide, as much material must be moved to obtain the uranium, which mostly occurs in low concentration. The energy needed to mine and mill uranium ore is about 2 GJ tonne^{-1}, and one tonne of ore might contain 1% uranium. It is useful to note that some uranium is obtained as a secondary process in the mining of some other metal.

Table 3.1 compares nuclear versus coal for specific item costs, for similar age and size electric generation plants on a US$ per megawatt-hour (US$10 MWh^{-1} = 1 cent kWh^{-1}). These costs are only indicative, as fuel costs and interest costs vary widely in time and depend in part on the global demand and speculative pressures.

Table 3.1. *Comparison of costs for nuclear and coal electric power generation.*

Item	Cost element	Nuclear US$ MWh^{-1}	Coal US$ MWh^{-1}
1	Fuel	5.0	11.0
2	Operating and maintenance – labour and materials	6.0	5.0
3	Pensions, insurance, taxes	1.0	1.0
4	Regulatory fees	1.0	0.1
5	Property taxes	2.0	2.0
6	Capital	9.0	9.0
7	Decommissioning and DOE waste costs	5.0	0.0
8	Administrative/overheads	1.0	1.0
Total		30.0	29.1

Reproduced from (www.nucleartourist.com/basics/costs.htm) in 2008.
DOE – US Department of Energy.

Security of the supply of fuel often influences the decision of whether to build nuclear or coal-fired plants. A national supply of coal can reduce fuel delivery costs. Table 3.1 treats carbon dioxide costs as an externality by not including them.

Decommissioning costs need to be allowed for when assessing the economics. A number of nuclear facilities have been dismantled, so such costs are known. As decommissioning costs come in the future, their present net cost can be discounted.

Electric and hydrogen cars

A significant fraction (about 20%) of the anthropogenic carbon dioxide comes from the transport sector. It is important, however, not to confuse the energy carrier with the source of energy. Electricity and hydrogen are energy carriers as there are no available supplies of these energies in the environment. If we converted to electric cars there would be no emissions of carbon dioxide at the car. However, if the electricity were generated from fossil fuels, there would be no saving in greenhouse gas generation except for possible changes in energy efficiency. If the electricity was generated by a zero-emission process, the 20% saving of the fossil-fuel generated CO_2 would be about $6 \times 0.2 = 1.2\,\mathrm{GtC\,yr^{-1}}$. The number of cars will be larger under scenario IS92a in 2030, but roughly the complete conversion of transport to zero-emission technologies would provide about 1.2/2 = 60% of the net emission reduction mentioned above as needed to be on a path to concentrations of 550 ppm. Whole lifecycle assessments are needed, since electrically powered transportation moves the problem to that of economically generating electricity without CO_2 emissions. There are a number of collateral benefits to do with improved air quality by not burning petroleum in the cities.

Growth in the number of vehicles is about 2% per annum according to Bekkeheien (1995), about the rate of increase of the GDP per person. At the same time the fuel efficiency of petroleum-fuelled cars has been improving.

The hybrid car provides a step change in fuel efficiency for city driving. Here a petroleum engine generates electricity under optimum load, and stores excess in a battery to drive the wheels through electric motors. At present the energy stored per unit weight of battery is low, limiting the ability to use grid-delivered energy. A new idea is to run the hybrid car's petrol engine when the car is not in use (say at night) to use the asset (the car) to generate revenue by providing electricity to the grid. This would be a distributed electricity generation system that may not help reduce carbon dioxide emissions.

Another interesting development is the electric bicycle, which allows the rider to either pedal or use the battery power. The batteries can be removed from the bicycle and recharged from the grid. Where there is no grid power they could be recharged from solar cells. About 20 million electric bicycles a year are sold in China at present, producing a total of 120 million already on the roads. Even wider acceptance could provide a way of using zero-emission power to reduce city air pollution and manage carbon dioxide.

Another alternative is the hydrogen car, where the energy carrier is hydrogen. Hydrogen presents some problems since, although it has the highest energy density by mass of all common fuels, with a lower heating value of about 120 MJ kg^{-1}, it has one of the lowest energy densities by volume at about 8 MJ l^{-1} as a liquid. Again hydrogen needs to be produced from a zero- (or low net)-emission process for the hydrogen car to contribute to the management of greenhouse gas concentration in the atmosphere. The distribution of energy is also an important consideration. Liquid petroleum has gained its ascendancy as a fuel for road transport because it is easy to store and move from its source to its point of consumption. In Chapter 2 we examined the efficiency of various ways of transporting energy. There is no retail distribution system in place for hydrogen.

Renewable energy

There is a group of electricity-generating technologies that are often termed renewables. They take this title from the fact that they derive their energy either directly from the sun or through the motion of the planets. Since we assume that these two sources will be available well beyond the foreseeable future, they are renewable sources of energy, in contrast with fossil fuels which we know to be finite. The fact that energy sources are available beyond the horizon of fossil fuels is of little concern to engineers. Engineers think in terms of present value (for a discounted cash flow discussion see Appendix 2). Thus energy resources in 300 years have

Table 3.2. *Renewable energy.*

Renewable energy	Description	Scale
Geothermal	Uses the earth-heat for direct application	Widely used in North America, Europe
Hydropower	Uses water to produce electricity	Extensively used worldwide; limited new capacity
Ocean (tide, wave)	Draws on energy of ocean waves and tides, and thermal energy stored in the ocean	Some full scale demonstrations
Solar	The direct use of the sun's energy to provide heat, light, hot water, electricity	Widely used in the world, but the scale for each project is small
Wind	Uses the energy in the wind to produce electricity	Used in many countries

negligible value today. Some say that currently people are discounting the future too heavily, and rational economics fail to conserve finite resources. In a quasi-*free market*, it is human sentiment that determines the discount rate, and this can be changed by popular perceptions.

Table 3.2 lists the technologies considered as renewable sources of energy. Geothermal energy extracts the heat stored in the earth and is less 'renewable' than some other sources of energy such as tidal energy, which is supplied from the kinetic energy in the heavenly bodies. The other energy technologies in Table 3.2 rely on solar energy. These sources of energy are a minor component of the world primary energy supply because of their present cost. The International Energy Agency (IEA) thinks that they will remain non-competitive in most circumstances while the cost of disposal of carbon dioxide (in the atmosphere) remains low. In favourable locations wind energy is the closest to being competitive with coal or nuclear.

What is important for the present discussion is that so-called renewable energies are generally clean sources of energy. Not only do they produce no carbon dioxide (though their manufacture might), but they are also free of other unpleasant chemical releases.

One of the difficulties of many renewable energy sources is that they are concentrated in regions away from cities, the location of much primary energy use. The cost of transporting the energy from its source to its use varies widely with the nature of the energy. Coal is easy to transport and this is a partial explanation for its widespread use. Natural gas is transported by pipeline at greater cost per unit of energy than coal but at lesser cost than electricity. This topic is treated in Chapter 2. It can be seen that renewable energies, once captured, must be converted to a transportable form, and those advocates of the hydrogen economy see this as the carrier of the future.

Wind power

The amount of wind power generation is growing fast. It seems to have been used first in the seventh century CE in Iran, for irrigation and corn grinding. The idea was soon being used in China (or was it the other way round)? Possibly it was transferred along the Silk Road, but in which direction?

The design has improved since the first windmills, and now a typical-sized unit is of 3.5 MW capacity. They are arranged in farms and then combined through a grid system. The grid system allows the power supply to be more uniform. When the wind is light in one area it is probable that it will be stronger in another (Figure 3.1).

The theoretically available power in the wind is given by

$$P = 0.5\rho \cdot U^3 \cdot A$$

where A is the blade swept area, ρ is the air density and U is the wind speed. The electricity generated is some fraction of this. It is clear that there is advantage for sites with large wind speeds, while locations with persistent winds are also favourable. We can consult wind maps to find sites with desirable wind characteristics. A criterion for a good site is mean annual wind velocity of greater than 7 m s^{-1}. Archer and Jacobson have recently published a global wind map in

Figure 3.1 See plate section for colour version. A wind farm. Image © Pedro Salaverría / www.shutterstock.com.

the *Journal of Geophysical Research*. See www.stanford.edu/group/efmh/winds/global_winds.html.

The operational and maintenance costs of wind power are low and there are no fuel costs. The price of producing the electricity is dominated by the cost of capital. Wind power is intermittent, and the capital costs reflect the fact that the average output is less than the maximum capacity.

Increasing the height of the turbine increases the power available a little, but by moving to the jet stream a big increase in the power is available. The idea is to power through a tether a helicopter-like device, to raise it some 10 km above the earth to be in the jet stream. Once in place, the rotors are able to generate power and transmit the electricity through the tether cable to the grid system.

Solar power

The solar energy incident on the earth's atmosphere is about 1360 W m^{-2}, but on the surface of the earth's atmosphere the average is a much smaller 340 W m^{-2}. A black coal power station might have a power output of 1000 MW, and this amount of energy could be collected from an area of 3 km by 1 km. However, nothing like 100% energy efficiency can be achieved. Using solar mirrors to increase the intensity of the solar energy, liquids can be heated to some 400 °C, and energy storage is possible with molten salt to provide electricity at night. Such facilities typically capture less than 30% of the energy that falls on the mirrors and have a capital cost of US$4000 per kW.

Commercial photovoltaic devices convert the incident thermal energy to electricity using the properties of silicon, and achieve efficiencies of less than 25%. However, cost per watt is probably a more important measure. Increased construction of solar cells has driven up the cost of silicon by a factor of 10 or so, but it is expected that extra production capacity will bring down the price of both silica and solar cells. The photovoltaic device needs energy for its manufacture and so one can talk of an energy payback period. At present it is a few years. As the efficiency of solar cells rises and the cost falls (the price has fallen 20% in the last decade), the cost of energy generation falls. Photovoltaic cells are finding increasing applications in remote locations where grid power is not readily available.

The solar chimney works by heating air under a large translucent roof which is vented to the atmosphere via a tall chimney (Figure 3.2). A test model operated in Spain has led to proposals for solar towers 1000 m tall. With an atmospheric lapse rate of 6.5 degrees per km, the solar collector roof design might be of 300 m radius with a solar collection area of 0.28×10^6 m^2. The solar energy incident on the roof is an average of 380 MW in Spain. Turbines are mounted at the base of the tower

Figure 3.2 See plate section for colour version. Solar chimney with its solar energy collecting roof made of green material.

to generate electricity. For some designs the temperature may reach 70 or 80 °C at the turbine entrance.

The heated air travels up the chimney, where it cools through the chimney walls. The chimney converts heat into kinetic energy. The pressure difference between the chimney base and ambient pressure at the outlet can be estimated from the density difference. This in turn depends upon the temperatures of the air at the inlet and at the top of the chimney. The pressure difference available to drive the turbine can be reduced by the friction loss in the chimney, the losses at the entrance and the exit kinetic energy loss. For a chimney of 10 m diameter and 200 m height, the power generated is about 50 kW on a sunny day in summer. If the height of the chimney were increased to about 600 m, the power output would have doubled. If still greater height were tried, the heat loss through the walls of the chimney would have been greater and the power generated less.

The cost of electricity from capital-intensive facilities such as solar cells and solar chimneys depends strongly on the interest rates and taxation rates (and allowable depreciation period). It is difficult to provide an estimate of the cost of capturing energy by capital-intensive facilities in the face of these uncertainties.

Geothermal power

The earth generates heat by the decay of radioactive isotopes which are present in the crust and core of the planet. There is a resultant average gradient of temperature of about 3 °C/100 m which supports a flux of order 65 mW m^{-2} from the continents, much smaller than the solar radiation incident on the earth. The flux of heat is not uniform and so there are opportunities to economically utilise this source of energy in certain locations. In the year 2000 there was about 8000 MW of electricity generated by geothermal power. As well, there was about twice as much energy used directly for heating.

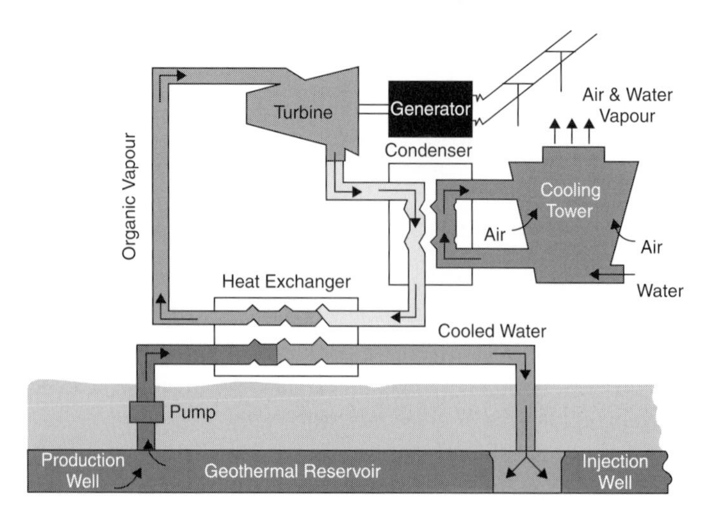

Figure 3.3 Typical geothermal energy plant of the binary type.

Geothermal power injects water into porous rock as in Figure 3.3, and extracts energy above ground by conventional means. The porosity of the rock can be increased by an order of magnitude by opening the natural fractures, by injecting high-pressure water into the geothermal zone. High temperatures at shallow depths are desirable to reduce drilling costs. In South Australia, test wells encountered temperatures of 250 °C in basalt rock at a depth of 4300 m. The additional capital cost of a geothermal plant comes from the drilling of wells needed to inject cold brine and gather, often through multiple wells, the heated brine. The pumping costs can be substantial and may consume some 20–30% of the electrical energy generated. Additionally, the geothermal zone cools some 40 °C over the 50-year lifetime of the plant. As the temperature of the rock is relatively low, and remembering Equation (2.1), the thermal efficiency of this system is low. Low efficiency in general means larger capital costs but has little impact on the operating costs as there are no fuel costs. The heat flow in the rock limits the power capacity of the system. A strong advantage of geothermal power is its continuous operation, making it suitable for baseload power generation.

The hot brine contains dissolved chemicals including carbon dioxide. These can be very corrosive and increase the cost of the heat exchanger if it must be constructed of stainless steel or titanium. While the hot brine can be used directly in lower capital-cost facilities such as atmospheric release turbines, there are the disadvantages of the release of carbon dioxide and the environmental impact of the discharged water. Almost all applications create a waste heat disposal problem that can lead to the requirement of the construction of a cooling lake or pipeline for disposal in the ocean.

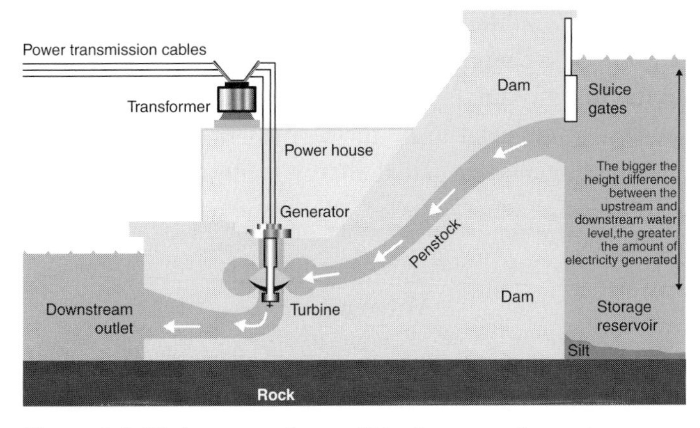

Figure 3.4 Hydropower. Image © hydropower.fireventura.com.

Performance/cost tradeoffs can lead to the cost of capital being up to 75% of the annual operating expense. A representative capital cost is US$3000 per kW and an electricity cost to the grid of US$0.1 per kWh.

Hydropower

Water stored at height has potential energy, and this is used to generate hydropower. The water is usually stored in a dam (Figure 3.4). It is dam failures that lead to the loss of life associated with this form of energy. Operation and maintenance (O&M) represents a small cost in hydropower, which is usually dominated by the need to service the capital used to create the civil engineering works.

Most of the sites suitable for large-scale hydroelectricity have already been exploited, but mini-hydro-projects are being built. The potential energy of unit volume of water h metres above the outlet tunnel is simply

$$PE = \rho_w \cdot g \cdot h$$

where g is gravity. A cubic metre of water stored 102 m above the outlet tunnel contains 1000 kJ of gravitational potential energy.

Tidal power

The flow of water and the rise and fall of the water surface in regions influenced by the tide can be harnessed to generate electricity. Because of tidal resonance, some regions such as the Bay of Fundy have a large tidal range, while other places, with narrow channels connecting larger water bodies, have high currents that can be exploited in a similar manner to wind power. The theoretically available power is the same as for the wind turbine except the density is that of water and about 1000 times greater. The Engineering Business, Ltd, reports that it has successfully installed its Stingray tidal stream generator in Yell Sound off the Shetland Isles, UK, and this is believed to be the world's first offshore installation of a full-scale tidal stream generator. Tidal energy is attractive because the tide is so predictable.

Wave power

Despite there being large amounts of energy available in ocean waves, this renewable technology has yet to be commercialised. One difficulty is that wave power devices must be able to survive the storms, while operating at much lower energy fluxes. The storm survivability drives up the initial capital cost. Surface gravity waves on the sea surface capture energy from the wind, and this energy can be converted to electricity. In deep water, a sinusoidal wave that has a height from the crest to the trough of H transports energy at a speed $c/2$ where c is the wave phase speed. Phase speed increases with the wave period and is $10\,\mathrm{m\,s^{-1}}$ for a wave of period 6.4 seconds. The energy flux or average power across a line normal to the phase speed is 1/16 of $\rho_w g H^2 c$. Thus, for a 2 m high wave with a phase speed of $10\,\mathrm{m\ s^{-1}}$, the power per metre of coastline (perpendicular to the direction of propagation) is 25 kW. A typical hybrid car engine has an output of 60 kW.

Costing of alternative energy

Costs are made up of three major factors, namely:

Capital costs
Fuel costs
Staffing and maintenance costs, O&M.

Each of these costs can be stated in terms of kilowatt-hours produced or capital costs in terms of rated kW. Bearing in mind that 'renewable' systems cannot produce electricity continuously, these figures in terms of rated kW need to be multiplied by a factor as high as five to obtain the average power.

The present capital costs per kilowatt are of the following order:

Coal, nuclear, wind	US$1000
Solar cells (domestic)	US$10 000

The current aim of the solar researchers is to reduce this cost to about US$1000 kW^{-1}, but that could well be a very optimistic objective. The decommissioning costs for nuclear plants occur far in the future and are not very significant on a present-day cost basis.

For comparison purposes it has been assumed that the capital costs would be provided by borrowed funds and paid back over 20 years, with an average interest rate over the period of 10% per annum. Then the so-called levelised annual capital cost (including repayment of the principal) is 0.118 of the total capital. The levelised cost is the present value of the total cost over the assets lifetime, expressed in equal annual payments. At a duty factor of 1 there are 8760 hours in a year.

The levelised capital costs (in US cents) per kilowatt-hour produced would then be as follows:

Coal and nuclear	1.4	(duty factor ~1)
Wind	4.2	(duty factor $\frac{1}{3}$)
Solar	60.0	(duty factor $\frac{1}{5}$)
Solar, if research objective can be met	6.0	

These figures do not take into account the capital costs of the backup equipment.

Fuel costs per kilowatt-hour in Australia are of the following order:

Coal – Victorian lignites	0.3*c*
Coal – black coal	1.1 to 1.5*c*
Gas	2.0*c*
Nuclear (including reprocessing, disposal or storage)	< 2.4*c* *(current USA figure)*
Solar and wind	0.0

Staffing and maintenance costs are very difficult to determine but are likely to be of the order of 1*c* to 2*c*, but could be higher for the newer technologies at the present time. Indeed we know that a large wind farm in California was closed down because of excessive maintenance costs, but the latest information is that later units have been improved considerably. These figures also penalise nuclear energy, as none of the others include waste disposal or site reclamation, and overall we know nuclear-generated power is being sold competitively with that generated from coal or gas.

Total operation costs then would be of the order of:

	Capital + fuel + O&M = total (US cents kWh^{-1})
Coal	1.4 + 1.1 + 1 = 3.5
Nuclear	1.4 + 2.4 + 1 = 4.8
Wind	4.2 + 0.0 + 1 = 5.2
Solar (current)	60 + 0.0 + 1 = 61.0
Solar (most optimistic)	6.0 + 0.0 + 1 = 7.0

While these figures are just order of magnitude, they show that nuclear power looks like a technical solution to the carbon dioxide emission problem that costs only a little more than black coal generation. Electricity will be able to replace much direct use of fossil fuels such as natural gas for household heating. Nuclear power has another advantage shared by wind power and the like, that price is insensitive to changes in fuel cost (increasing in the future). However, nuclear power is politically unpopular.

Energy storage

The above renewable energy sources are intermittent and they would be much more attractive if the energy collected during the high-wind periods or the daylight hours could be stored until demand was high. In the case of fossil fuels, the fuel is not burned until the energy is required. Fossil fuels are efficient stores of energy. Renewables, however, have to capture energy immediately.

There are three promising methods of storing energy: batteries, capacitors or heat storing devices. Batteries rely on reversible chemical reactions such as lead acid, nickel lithium ion or lithium ion polymer. All have limited lifetimes and relatively high cost per kWh of energy storage. Lead acid batteries are typically US$1500 per kWh. Flywheels are a form of capacitor that have good energy density and cost some US$300 per kW.

One promising heat storage system employs molten salts. A mixture of 60% $NaNO_3$ and 40% KNO_3 is molten at atmospheric pressure and about 220 °C (430 °F). Still another system of large capacity is pumped hydro-storage. The capital cost of storage able to deliver 1 kW of power is of order US$1000. Since pumped hydro-storage would only be used for a fraction of the time, the storage is a rather expensive option.

Carbon capture and storage

If the carbon dioxide generated from fossil fuel burning, such as in electric power stations, steel mills or from carbonate in cement manufacture, is not emitted to the atmosphere at the plant but rather disposed of in geological structures or in the deep ocean, then a low-emission (near-zero) facility can be operated. The technology is established, but the cost remains too high to make it economic at the current (2009) prices of carbon credits. This cost is dominated by the cost of separating the low (less than 15%) concentration of carbon dioxide in typical flue gases from air burning boilers. Since the price of carbon credits may rise in the future, there are a number of plans to demonstrate, on a commercial scale, the generation of electricity from fossil fuels and then sequestration of the carbon below the earth surface. This topic is treated in greater detail in Chapter 6.

Next we turn to look at ways to avoid the capture cost and produce a stream of nearly pure carbon dioxide.

Oxygen economy

Since flue gas from fossil-fuel power stations contains only about 15% CO_2, capture turns out to be an expensive technology, particularly as coal-burning power stations often produce significant sulfur emissions. An alternative approach is to separate the O_2 in the air from the N_2 before combustion. Oxygen represents about 20% of air and costs much the same to separate from air as carbon dioxide.

With oxyfiring, rather than

$$C + O_2 + 4N_2 \rightarrow CO_2 + 4N_2$$

one can use

$$C + O_2 \rightarrow CO_2$$

Sometimes there is benefit in including water

$$2C + 2H_2O + O_2 \rightarrow 2CO_2 + 2H_2$$

where the water controls the combustion temperature. Separation of water from the exhaust gas stream is simple, to give a sequestration-ready stream of carbon dioxide. By not passing the nitrogen through the combustion system, its size can be reduced and the temperature of combustion increased. The smaller gas flow reduces the size of the gasifier, while the increased temperature leads to more complete burning of the coal. By using a gas turbine with a steam turbine, termed a combined cycle, a higher inlet temperature is possible while retaining the traditional exit temperature, and this improves the Carnot efficiency, Equation (2.1).

Table 3.3. *Electricity cost for 400 MW plant operating on oxygen with gas turbines or air-breathing combined cycle technologies in year 2000 US dollars.*

Plant operating factors	Oxygen cycle				Air combined cycle		
Turbine technology[a]	Current	Near-term	Advanced		Current technology		
Thermal efficiency (%) (with syngas plant)	32	37	43	44	46	37	38
Capital cost (US$ kW_e^{-1})	1425	1525	1525	1365	1480	2110	2030
Coal cost (US$ GJ^{-1} (LHV))	1.70				1.70		
Emissions of NO_x (kg MWh^{-1})	0.00				0.03	0.04	0.04
Emissions of CO_2 (kg MWh^{-1})	*0.00*				*745*	*139*	*135*
Capital unit cost (US$ kWh^{-1})	0.031	0.033	0.033	0.029	0.032	0.045	0.044
Fuel cost (US$ kWh^{-1})	0.019	0.017	0.014	0.014	0.013	0.017	0.016
Maintenance cost (US$ kWh^{-1})	0.008	0.007	0.007	0.006	0.007	0.009	0.009
Cost of electricity (US$ kWh^{-1})	*0.058*	*0.057*	*0.054*	*0.050*	*0.052*	*0.071*	*0.069*

www.cleanenergysystems.com/.
kW_e = kilowatts of electricity.
[a] Higher turbine temperatures give better efficiency.

These 'oxygen' cycles can take many forms. An attractive proposal, put forward by Anderson *et al.* (1998) and by Bilger (1999), involves obtaining oxygen by cryogenic separation. Table 3.3 compares the costs of using oxyfiring with a current-technology gas turbine, a near-term improvement and two advanced gas turbines. These can be compared with examples of three current turbines using air combustion. Two of these cases include carbon capture using organic solvents (monoethanolamine – MEA). When the disposal of carbon dioxide is considered, oxyfiring is cheaper.

Notice that the cost of electricity is greater than in the previous calculation, with the cheapest gasification option being US$0.052 per kWh. (Most of this difference comes from different estimates of fuel costs.) The lower capital costs projected for the oxygen cycle in Table 3.3, and the increased efficiency after capture of carbon dioxide is considered, makes it a preferred technology for the future. However an early adaptor of this technology can expect to incur a cost penalty, and so far this is inhibiting the introduction of the technology.

Energy from biomass

By using vegetable matter to provide energy, the **net** emission of carbon dioxide can be small. Plants take carbon dioxide from the air to grow, and the burning of the biomass returns the carbon to the air. Firewood is an example of a zero-net-emission fuel that is widely used. Recently, specially grown crops are being used to make biofuels. Corn, for example, can be converted to alcohol and used as a partial petroleum substitute. As you need to supply nitrogen and other nutrients that are taken from the soil for the process to be sustainable, these are often produced by an industrial process that needs energy. Farm machinery also uses energy and at present releases carbon dioxide to the atmosphere. There are some estimates that a litre of corn alcohol requires 0.8 litres of fossil fuel to produce. Another important issue is the opportunity cost of not providing food. Already the global rise in food costs is being attributed to the production of alcohol for use as a fuel. It is the poor that suffer the most from rising food costs. Cellulosic alcohol seems preferable as it is made from the non-edible parts of plants, but presumably costs more to produce at present.

Biodiesel can be produced from vegetable oil or animal fat by a well-understood chemical process. Soybeans are the main source of vegetable oil at present and produce typically 700 litres per hectare. Production will be limited by the supply of agricultural land and water.

The prospect of producing biodiesel from marine phytoplankton has received some attention. Phytoplankton grow in the sunlit upper ocean and contain up to 75% oil on a dry weight basis. By using a marine region one does not suffer the same opportunity cost as in producing biomass on land which has many uses such as agriculture or mining. Phytoplankton get carbon from the environment and use solar radiation as an energy source. A method of economically harvesting the phytoplankton has yet to be invented.

Lifecycle accounting

Many governments are mandating that a certain fraction of electricity carried by the electric grid system comes from renewables. The reasons given for this include the fact that fossil fuels are finite and will eventually be exhausted; the free market in fossil fuels puts a value on this future scarcity. One can argue that the present pricing of coal is too low and is a case of market failure to adequately consider the future, but one would need to produce evidence of this. It is economically rational to use fossil fuels while they provide power at competitive prices with the future scarcity factored in. A second reason for mandating renewables is that they appear to emit no greenhouse gas. Now we need to see at what cost comes the saving in CO_2 emissions.

We need to consider not just the emission of greenhouse gas during the operation of a process but also the carbon dioxide generated by its manufacture, transportation, maintenance and eventual disposal. This is known as lifecycle costing.

Exercises

Exercise 3.1 Calculate the increase in the cost of electricity due to capture of the carbon dioxide from both a coal-fired power station and a natural gas station. Assume capture can be accomplished at US$40 per tonne of carbon dioxide (capital and operating cost) and that the process uses 3000 MJ of energy for each tonne of captured CO_2. Assume natural gas is available at US$4 per GJ. What is the expected change in electricity price at the bus bar (wholesale) due to carbon capture? Sequestration will be an additional cost. How does this compare with fluctuations in oil prices? Would the impact on the global economy be of the same order as the oil price rise?

Exercise 3.2 It has been proposed that by the year 2050 energy will be transported by hydrogen. Explain why hydrogen is considered more attractive than our traditional energy transport systems such as electricity.

4
Geoengineering the climate

In the previous chapter we considered approaches to providing energy with near-zero emissions of carbon dioxide. While this may be technically possible, there are impediments to the adoption of these concepts. Some are political, some are economic and some are resistance to change. The alternative approach is to accept the rise in carbon dioxide concentration in the atmosphere due to continuing emissions of carbon dioxide, and to modify some other components of the climate system to maintain a desirable climate. This is known as geoengineering – engineering on a global scale. It implies exerting control over nature, a concept that comes more naturally to engineers than to others with different cultures.

Five hundred years ago, humans had made only a small dint on the global ecosystem. The land biomass was presumably in steady state, so that on the average it neither stored carbon, nor released it to the atmosphere. Then came land clearing for agriculture, with the consequent release of carbon dioxide. As the CO_2 level started to rise with the Industrial Revolution, carbon flowed from the atmosphere to the sea because of Henry's Law. The ocean sink is currently estimated at 1–2 $GtC\,yr^{-1}$. One way to make this estimate is to measure the carbon dioxide partial pressure difference between the atmosphere and the ocean surface layer and use this in a flux calculation. This topic is discussed in Chapter 5.

What is geoengineering?

The intentional large-scale manipulation of the climate could be termed a geoengineering project (Keith, 2000). It could take the form of planting vast forests to temporarily sequester atmospheric carbon, or it could involve the spreading of particles in the upper atmosphere to reflect more sunlight. There are other localised schemes such as damming the Mediterranean or removing the Arctic ice to warm Russia (Rusin and Flit, 1960). It has been suggested by an anonymous author (Anonymous, 1963) that the ice cover of the Arctic, once removed, would never be re-established. Weather modification (cloud seeding) is an early attempt at geoengineering.

The majority of people in the world will not have heard of geoengineering because they are focused on making an adequate living for the family. Amongst the privileged with the resources and education to consider environmental issues, geoengineering is a concept that produces apprehension. Jamieson (1996) points out that people are relatively unfamiliar with geoengineering technologies, and this heightens their anxiety. Even though we are unintentionally changing the climate by releasing carbon dioxide into the atmosphere, the impact of intentional intervention is viewed differently, from a moral standpoint, from interventions that occur inadvertently – errors of commission rather than errors of omission. Whether this is a sensible differentiation is a moot question. It could be considered a human foible that a disaster, incurred while trying to reduce distress by engineering the climate, would be judged more harshly than failing to take action and then suffering the undesirable consequences of increasing emissions of greenhouse gases. The undesirable impacts of irrigation, such as the salination of surface soils, are often cited with disapproval. Humans embarked on irrigation to increase agricultural yield before all the consequences were recognised. While there have been some setbacks, overall, innovations have contributed to a doubling of the agricultural yield in the last 50 years. Those changes that, implemented in the face of uncertainty, worked out exactly as planned, were applauded. Those that did not were judged harshly in hindsight.

One engineering approach to uncertainty is to implement schemes in ever-increasing scale while monitoring the consequences. Constructing a pilot plant before full-scale implementation is a time-honoured approach. In agriculture, people have faith that human ingenuity can keep up with various problems (such as new pests) that novel approaches are bound to produce. The outcome has been spectacular, with agriculture feeding more than six billion people.

The magnitude of engineering on a global scale, implied by the term geoengineering, also worries people. While the public can countenance a small change from the *status quo*, a bold vision of the future often raises alarm. We are inclined to protect our current situation by treating geoengineering as science fiction or by refusing to examine the issues raised by such future scenarios. For instance, in Australia, an early action of a newly elected Conservative Government was to disband the Commission for the Future!

Keith (2000) addresses the concept of moral hazard. This is the issue of taking actions where you pass the risks involved to another party rather than bear the full consequences yourself. Would the knowledge of the existence of a future viable geoengineering solution to an environmental problem weaken the will to mitigate climate change now? This is like the insurance question. It encourages others to take risks that are individually advantageous but are not optimum for the larger group of insurers. As an example, the knowledge that Ocean Nourishment (discussed in Chapter 5) is a low-cost way to produce a carbon sink might encourage the postponement of reducing emissions of carbon dioxide. Then if the carbon sink produced was short lived, it would have allowed some to continue to place waste carbon dioxide in the atmosphere to the disadvantage of others. A moral hazard, it is claimed, because the reckless action might disadvantage others! This line of thinking is related to the uncertainty involved. If producing sinks is cheaper than avoiding emissions, and if there is low uncertainty that the sink lifetime may be misunderstood, the sink choice is justified and the moral hazard is avoided.

It may be this moral hazard concept that leads a number of environmental lobby organisations such as Greenpeace (Johnston *et al.*, 1999) to oppose the use of what they term *planetary engineering* to combat climate change. They fear it may lull people into false confidence and cause policy makers to avoid taking more painful action now.

Policy makers have a difficult time resolving the conflict between environmental interest and those wishing to dispose of carbon dioxide in the atmosphere. There is an immediate reward in the lower cost of carbon dioxide emission to the atmosphere while the benefits to the environment of slower climate change will mostly accrue to a later generation. How can a value be put on benefits such as the protection of diverse assemblages of creatures? What is the value of biodiversity?

Climate geoengineering techniques

Let us defer to future chapters the discussion of strategies to capture carbon dioxide and store it away from the atmosphere, such as storing carbon in trees, in geological structures and in the ocean. In this chapter, we will focus on concepts that control the radiative forcing of the earth. Solar energy reaching the top of the atmosphere is nearly constant at $1367\,W\,m^{-2}$. It varies only a few $W\,m^{-2}$ from decade to decade. By placing objects between the earth and the sun, as in a solar eclipse, the energy falling on the top of the atmosphere could be reduced. The amount of short-wave radiation reflected back to space by the atmosphere depends upon the clouds and upon scattering by aerosols, and these could be modified. Of that energy that reaches the surface of the earth, some is reflected back into space and this varies seasonally as the colour of the earth changes. The fraction of the short-wave solar radiation incident on the top of the atmosphere that is reflected into space is known as the planetary albedo, and reducing the earth's albedo would change the radiation balance of the earth. Increasing the albedo will lower the temperature of the planet.

Geoengineering the climate does not seem to be considered in international law, and the UNFCCC remains silent on the topic of modifying the short-wave radiation. International agreements can be expected to be difficult to achieve because modifying the climate is certain to create winners and losers. While compensation between nations could be considered, this would be a very big development in international values which at present are firmly rooted in sovereign states. Concepts such as controlling transboundary 'pollution' are examples of individual nations being concerned about the disadvantage they might be visiting on their neighbour state. Sulfur particles generated in one country and falling as acid rain on another country is already causing conflict in Asia, and public opinion is likely to play a more important role in the future.

Climate engineering involves a high degree of uncertainty. There is little experience to fall back on. This new form of engineering will need to develop strategies to manage adverse impacts or even reverse procedures that are not beneficial.

Solar radiation

The temperature of the earth is related to the solar radiation that falls on the face of the earth exposed to the sun. In the region that is experiencing night there is no solar radiation. For the last 20 years, direct measurement of the solar irradiance has been made. Before this period, solar irradiance must be estimated from a model of irradiance in terms of solar activity indices such as sunspots. Learn *et al.* (1995) have constructed such an index from records of sunspot occurrences back to 1610. The Chinese Empire was very interested in such things. Needham (1959) noted the correlation between hot and cold centuries (sixth to sixteenth) and sunspot frequency in Chinese records.

Climate change sceptics consider that the evidence for greenhouse gas warming of the earth is weak. They speculate that the changes in solar radiation or other natural causes might be responsible for the increase in temperature reported by the IPCC.

Let us look at the radiation balance of the earth. The solar radiation on the top of the atmosphere varies slightly about a present value of $1367\,W\,m^{-2}$, mentioned in Chapter 1. This is not the average radiation per square metre of the earth's surface, because any point on the earth is in darkness some of the time, and because the earth is a sphere most surfaces are not normal to the sun's radiation. This can be seen from recognising that the solar radiation intersects a disk of area πr^2, while the surface of the earth is an area of $4\pi r^2$. Here r is the radius of the earth.

The changes in the radiation reaching the top of the atmosphere are mostly due to sunspot activity. Variations in the earth's distance from the sun change on a much longer timescale. Reproduced from a World Wide Web page in Figure 4.1 is an estimate of the changes in solar radiation based on the last 400 years of observations of

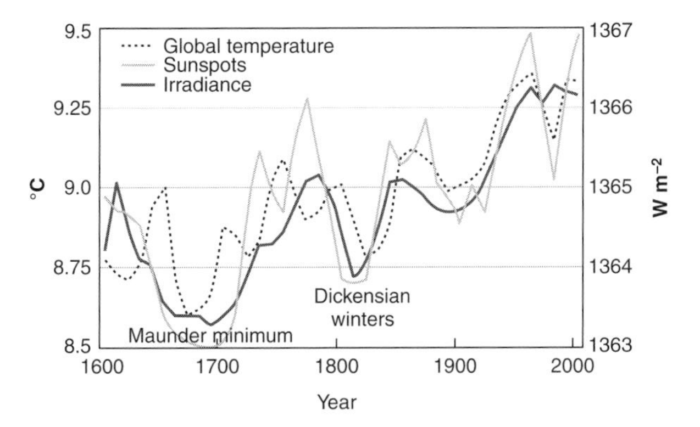

Figure 4.1 Correlation between global temperature and solar activity, CE 1600–2000 (irradiation from Fligge and Solanki, 2000 using a model based on sunspots. These are not direct irradiation measurements). This graph has been taken directly from the World Wide Web and is not verified.

sunspot activity. Variations of only a few tenths of a per cent are suggested for the last 100 years, and the radiation trend is upwards. The direct effect of this increase (that is, ignoring climate feedback) would be a rise in the earth's temperature. If this trend was reversed, it could be expected to lead to a potential cooling of the earth. Those that claim that most of the observed temperature rise at the earth's surface is related to fluctuation in solar radiation need to hypothesise an amplification process. It is not clear what this might be. If there were both a large decrease in radiation and an increase in greenhouse gas concentration in the future, these two influences might offset each other.

The solar radiation reaching the earth's surface is fluctuating, and for the last 50 years has been decreasing, a phenomenon known as global dimming. Stanhill (2007) reports decreases of order $20\,W\,m^{-2}$ in some places. The reason for this change is not clear, although aerosols are a prime suspect.

The radiation balance of the earth could be changed by providing a reflecting shield in space to reduce the incoming short-wave radiation. This is the concept of placing a sunshade between the sun and the earth. If this sunshade diverted $17.6\,W\,m^{-2}$ of incoming solar energy, the average radiation per square metre on the earth's surface would be reduced by $4.4\,W\,m^{-2}$. Thus a sunshade needs to deflect this amount of radiation to balance the increase in heat load in 2060 from the extra carbon dioxide expected in a *business as usual* scenario. The L1 point is a position where the combined gravity forces of the sun and the earth keep a third object in a constant relative position. The radius of such an orbit around the earth is about 1% of the distance between the sun and earth. If we were to place an opaque object between the earth and the sun at the L1 position, what diameter would it need to be to reduce the radiation intersecting the earth? For an opaque object to stay at the L1 point, the radiation pressure on the object needs to be considered. The cost of this mitigation idea appears to be much larger than others and is not pursued further.

Albedo changes

Of that solar energy which intersects the planet, the fraction reflected back to space is known as the earth's albedo (Figure 4.2). The current planetary albedo of the earth is about 30%, while the albedo of a black body is zero. The planetary albedo is mostly due to scattering of solar radiation by the clouds, but aerosols also influence the albedo, as does the surface reflectance of the earth. Reflectance is the fraction of the incident energy reflected at a surface.

By increasing the atmospheric aerosols one could reduce the solar radiation that penetrates to the surface. Volcanoes, when they erupt, often release large clouds of aerosols. The 1991 eruption of Mt Pinatubo released sulfur dioxide into the atmosphere, which reacted with the water to form aerosols of sulfuric acid. It is generally believed to have caused a global temperature decrease of 0.5 °C.

Aerosol changes

Keith and Dowlatabadi (1992) discussed the idea of injecting SO_2 into the stratosphere to mimic the actions of large volcanic eruptions. Possibly injecting 10 Mt

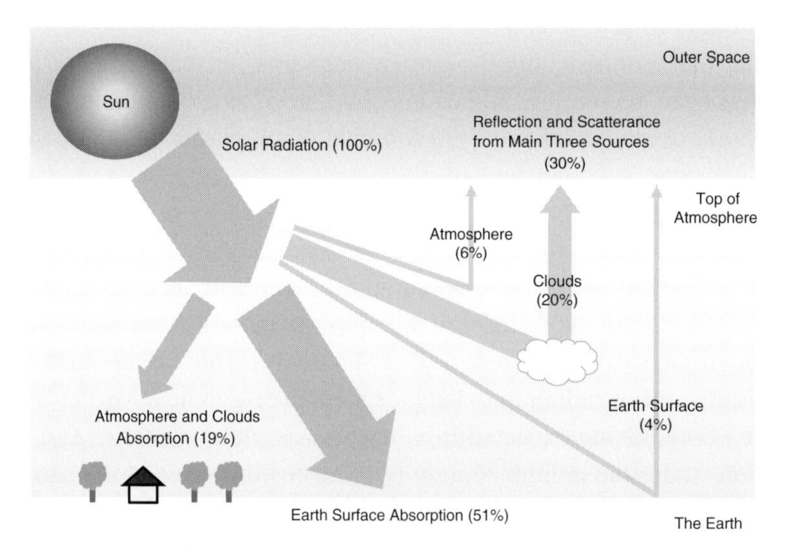

Figure 4.2 Albedo terms showing the short-wave energy paths. The earth's surface reflects 4% of the short-wave radiation and absorbs 51%. Much of this 51% is reradiated as long-wave radiation.

per year of sulfur particles would counteract the heating involved in raising the concentration of CO_2 from 280 to 560 ppm. This would need to be done each year, but the amount injected in the near term to achieve the equivalent temperature to 280 ppm would be less.

Unlike switching to zero-emission technologies, the management of atmospheric aerosols needs to be continued for the lifetime that the carbon dioxide is in the atmosphere. It is difficult to compare the cost of a one-off expenditure with one that must be continued for ever to obtain the same result. If one wishes to compare *present cost* (see Appendix 2), then a period and a discount rate must be agreed. A common discount rate is 5% pa. If the period is extended from 100 years to 200 years, it increases the present cost by less than 1%. Thus we will use present cost for 100 years as an indicator of the equivalent cost per tonne of CO_2 mitigated.

This natural phenomenon of particle release from volcanoes also provided the motivation for Paul Crutzen (2006) to revisit the idea of aerosol injection as a geoengineering solution for greenhouse gas warming. If the reduction in radiation was large enough it could stop an increase of the earth's surface temperature. Possibly the present intercontinental aircraft flights could be used to distribute particles which are expected to have a large residence time (of order one year). Condensation trails of aircraft are already having an impact on the climate. The estimated cost is US$25 billion per year to compensate for $0.75\,W\,m^{-2}$ of greenhouse gas heating (Crutzen, 2006). The radiation forcing in the year 2000 was about twice the above value, and represents the greenhouse gas heating of about 200 GtC of emissions since the Industrial Revolution. Thus, to reflect from the stratosphere this amount of radiation for 100 years, the present cost would be about 20 times the yearly value; that is, 20 × US$50 billion = US$1000 billion. This equates to a cost of US$5 per tonne of carbon dioxide compensation. Such a present cost is low compared with alternatives such as switching to zero-emission technologies, and could be implemented quite quickly.

If the present cost was set aside in a trust fund and used for 100 years of particle refreshment, this would go some way to answering the criticism of geoengineering that it runs the risk of sudden cessation exposing the earth to a rapid rise in temperature.

It is interesting to note that the world is moving in the opposite direction. Efforts in air pollution reduction that lead to a decrease in SO_2 emissions will lead to a lower reflectance and thus a heating of the earth. The benefits to human health are believed to outweigh the penalties of climate change.

Another approach is to make the clouds whiter. If the albedo of low-level maritime clouds were increased, the earth would be expected to be cooler. This could be achieved by increasing the number of saltwater droplets in the lower atmosphere

Figure 4.3 See plate section for colour version. An artist's concept of the ship to provide seawater droplets to form low-level marine clouds. Rate of water injection 1 kg sec^{-1}. Image John MacNeill.

that evaporate and leave behind a tiny particle of sea salt. If these condensation nuclei reduce the cloud particle size, they decrease the precipitation efficiency of the cloud and so increase the cloud lifetime. Smaller particles also increase the optical thickness of the cloud. Water vapour condenses on these salt particles and increases the cloud's reflectivity. Figure 4.3 shows an artist's impression on how this might be achieved using the wind as a source of energy for the vessel. High-pressure seawater jets would produce small drops which would be injected into the turbulent marine boundary layer.

In order to reflect the same amount of heat as trapped by 0.5 Gt of carbon dioxide, Salter *et al.* (2008) estimate that one would need 50 of the vessels shown in Figure 4.3. Each vessel might cost US$2M, with a lifetime of 20 years, and they would need to be maintained forever. If we use a discount rate of 5% and maintain the boats for 100 years, we can calculate the present value cost. The first boat costs US$2M. The second boat purchased in 20 years time has a present value of US$$2(1 - 0.05)^{20}$M and so on. The present value cost of providing one boat is US$3M, or US$150M for 50 boats maintained for 100 years. This would

Table 4.1. *Reflectance (%).*

Material	Range	Typical
Oceans	5–20	8
Moist soils	5–20	12
Dry soils, desert	20–40	32
Short, green vegetation	10–20	17
Dry vegetation	20–30	25
Forest	10–25	15
Sea ice	25–40	30
Fresh snow	60–90	75
Asphalt pavement	5–10	7
Concrete pavement	15–35	20

be equivalent to a reduced radiative loading of 500 × 10^6 tonnes of CO_2; that is, less than US$0.3 per tonne of CO_2. This is so much less than many other alternatives that more careful costing is not necessary. This concept seems easy to stop and clouds would quickly return to their usual state. The reduced radiation would not be uniform over the globe and might bring changes in the weather patterns. As well, it does not address the problem of ocean acidification.

Reflectivity of the earth's surface

Of the 55% of incoming radiation that strikes the earth's surface, about 4% of the incoming radiation is reflected, as shown in Figure 4.2. The short-wave reflectance of the surface of the earth varies between different land cover types as shown in Table 4.1, and it varies seasonally. As 70% of the earth's surface is sea, the average reflectance needs to be near the reflectance of the sea, and we see from Table 4.1 that it is typically 8% of the radiation that strikes the surface of the earth or 4% of the radiation at the top of the atmosphere. When Brazilian ranchers cut down dark, tropical rainforest trees to replace them with even darker soil in order to grow crops, the average temperature of the area allegedly increases by about 2 °C year-round. The heat island effect around cities is partially due to a decrease in reflectance of the buildings and streets over that of green fields. On the other hand, studies in the Amazon basin have demonstrated that a conversion of rainforest (typically 15% reflectance) to pasture (typically 17% for green, 25% for dry) results in an increase of surface albedo from ~12–14% to 24–26%. This is a temperature-reducing change.

Global scale changes due to land clearance for pastures are arguably the greatest direct influence of human activity on the earth's terrestrial system. The greatest human-induced reflectance (albedo) changes are occurring in the tropics. These changes of land use for agriculture are varying the reflectivity and in turn the planetary albedo.

The short-wave energy reradiated into space is given by

$$radiation = 188 \times R$$

where R is the reflectance (albedo), and the average radiation on the surface per square metre of earth is about 1376 × 0.55 / 4 = 188 W m^{-2}. Thus a change from 8% to 18.6 % of surface albedo leads to a decrease in heat of 20 W m^{-2}. If this change were made over 30% of the land (30% of 30% = 9% of the earth's surface), the radiative forcing would be 1.8 watts per square metre of the earth's surface. Radiative forcing is a measure of climatic change of the energy balance of the earth. The variation in incoming solar radiation shown in Figure 4.1 was about 4 W m^{-2} (1 W m^{-2} averaged over the whole earth surface). How does this compare with the energy trapped by a change in the concentration of carbon dioxide in the atmosphere?

Radiative forcing and surface temperature

As well as the short-wave radiation shown in Figure 4.2, the surface of the earth radiates long-wave radiation to space and, as mentioned previously, some of this is absorbed in the atmosphere by greenhouse gases. We can estimate the amount absorbed by calculating the terms in a heat balance. To match the current temperature of the earth, one needs to assume a long-wave absorption factor, G, of 0.4.

Houghton *et al.* (1990) provide some expressions to relate the radiative forcing to the change in atmospheric concentration of carbon dioxide. This is the change in the forcing per area of the earth's surface, not the change in the incident radiation. When ΔF is measured in W m^{-2}, then

$$\Delta F = 6.3 \ln\left(C / C_0\right)$$

This logarithmic expression arises because the strongest absorption lines are saturated.

To illustrate, imagine that the concentration of carbon dioxide changes from 370 to 400 ppm. The ΔF becomes $6.3 \times 0.078 = 0.5$ W m^{-2}, while a doubling of CO_2 (from pre-industrial values) makes the radiative forcing increase by 4.4 W m^{-2}. Note that this expression includes the 'positive feedback' of the climate system.

The various elements that go into changing the radiative forcing are illustrated in Figure 4.4.

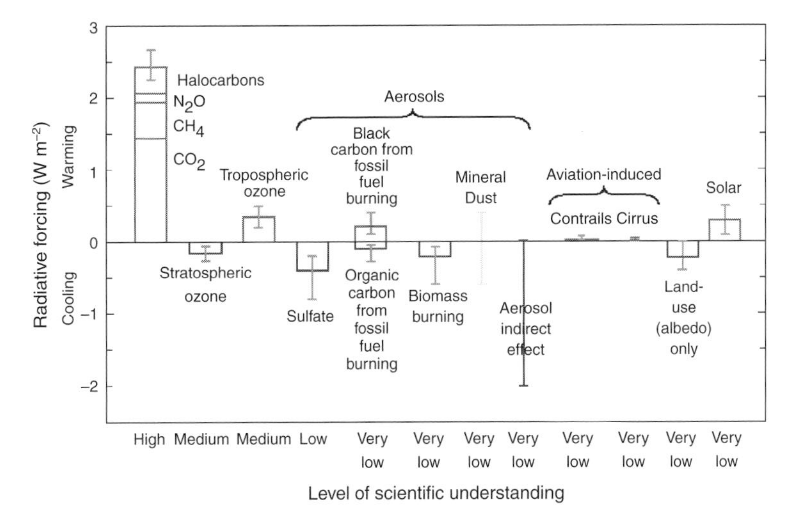

Figure 4.4 See plate section for colour version. The global mean radiative forcing of the climate system for the year 2000, relative to 1750, reproduced from Houghton *et al.* (2001).

Modifying the atmospheric water vapour

Water vapour is an important greenhouse gas. About 70 W m^{-2} of heat from the sun evaporates surface water on the earth, and this condenses into clouds in the atmosphere, releasing heat that drives the atmospheric circulation. It is thought that the concentration of water vapour in the atmosphere should be determined within the climate system as a response to changes in radiative forcing. This is why you do not see water vapour listed as an anthropogenic greenhouse gas. Could the water vapour level be changed by purposeful intervention? To answer such a question we would need an improved understanding of the global water cycle. Since about 70% of the earth's surface is ocean, the impact on the water vapour emission from the land into the air will be small. It is hard for an engineer to imagine changing the flux of water from the sea. However, there are some that can imagine spraying water into the atmosphere from boats to increase evaporation. This would be a much bigger activity than providing cloud nuclei as discussed above. Might increased irrigation be changing the water vapour concentration?

The long-wave planetary radiation is strongly absorbed by water vapour. Water vapour and other greenhouse gases reduce the long-wave radiation to space by about 146 W m^{-2} on clear days and another 30 W m^{-2} in cloudy regions. This is large compared with the extra radiation forcing of carbon dioxide, as water vapour contributes some 60% to greenhouse absorption while carbon dioxide contributes about 26%. Changing the water vapour will change the clouds. Increasing water vapour absorbs more long-wavelength energy, but the increasing cloud cover reflects more solar radiation. This is an area of uncertainty in climate modelling and it is not clear that we can achieve much climate intervention even if we could change the flux of water into the atmosphere.

Summary of geoengineering opportunities

This chapter has looked at the options to accept the increased concentration of carbon dioxide in the atmosphere and to change the thermal balance of the earth by manipulating the other terms in the heat balance to maintain the earth's surface temperature. Such options do not address other issues such as changed vegetation due to carbon fertilisation and ocean acidification. Some of the changes due to increased carbon dioxide will be beneficial in the same way as a temperature increase will improve conditions in some high-latitude areas. Forestry, for example, is expected to benefit from increased carbon dioxide concentration.

While the costs suggested for the concepts discussed in this chapter are very preliminary, they can be placed in ascending order. Injecting salt droplets into the marine boundary layer using the wind as an energy source appears to have the lowest cost, and this cost is a fraction of a dollar per tonne of CO_2 avoided. Injecting sulfur into the stratosphere holds out the promise of costing a few US dollars per tonne every year. The cost of these two concepts appears to be modest, and they can quickly be reversed if needed. The sunshade in space appears the dearest and will be hard to reverse.

All these concepts need research and development. Few people advocate geoengineering the climate today, but there is an increasing number who advocate research and development to prepare for the types of geoengineering discussed in this chapter in the event of a catastrophic climate change. Geoengineering would be likely to have unexpected consequences, but if the alternative is the present unplanned experiment which is now being conducted via the emission of carbon dioxide from power stations, motor cars and cement production facilities, then geoengineering is certainly worth exploring.

The next two chapters deal with creating sinks for the carbon dioxide once it is liberated. This involves atmospheric capture of carbon dioxide and its storage away from the air. If the sinks were of global scale they could be classified as geoengineering projects. Creating forest in the Australian desert (if a practical way to overcome the lack of water were found) could be an example, as would nourishing vast sweeps of the ocean.

Exercises

Exercise 4.1 The area of the Australian desert is about 3.8×10^{10} m^2, while the surface area of the globe is 510×10^{14} m^2. Examine the proposal that the albedo of the earth could be changed to counter the heating due to the amount of carbon dioxide released over that taken up by the ocean and the land. Assume that the carbon in the atmosphere increases about 2 Gt yr^{-1}. Once you have determined the change in albedo, estimate the present albedo and suggest how a change might be achieved.

Exercise 4.2 Assume that, in 2000, the fate of the 6 Gt yr^{-1} of anthropogenic carbon dioxide released was that 2 Gt stayed in the atmosphere, and 2 Gt went to the ocean and 2 Gt went to the land. If emissions increased to 8 Gt yr^{-1} or 10 Gt yr^{-1}, how does the radiation forcing change? Can you calculate the change in the earth's temperature by assuming the Boltzmann constant for black body radiation is 5.7×10^{-8} W m^{-2} K^{-4}?

Exercise 4.3 There is a strong visual correlation between the solar radiation and the global temperature in Figure 4.1. This would seem to explain the changes in temperature without recourse to greenhouse gas warming. Comment.

5
Ocean sequestration

The threat of rapid climate change due to greenhouse gas build-up in the atmosphere is the result of the sources of greenhouse gas to the atmosphere exceeding the sinks. The transformation of CO_2 in the atmosphere is very slow. In the previous chapter we considered how to change the radiative balance by adjusting the energy flux provided by the sun. We should now turn to examining the issues in using the ocean as a sink of carbon. Already the total mobile carbon in the oceanic waters is of the order of 40 000 GtC, much greater than the 2200 GtC on the land. The carbon in the ocean is stored mostly as bicarbonate, and the total dissolved inorganic carbon has a concentration of order 2000 $\mu mol\ kg^{-1}$.

Carbon is cycled by the marine planktonic ecosystem. Houghton *et al.* (1996), in the second assessment report of the Intergovernmental Panel on Climate Change, concluded that if there were changes in the oceanic plankton, there is a large potential for the biological pump to influence CO_2 concentrations in the atmosphere.

Carbon is also cycling in and out of the ocean as a result of the vertical circulation in the large ocean basins bringing water to the surface with a different partial pressure to that of the atmosphere above.

Ocean atmosphere exchange

Carbon dioxide flows across the air–sea boundary driven by the partial pressure of carbon dioxide. When the partial pressure in the air is greater than in the water (typically the case near the poles), the flux of carbon is into the water, while near the equator the partial pressure is typically greater in the water than in the air, leading to outgassing. The partial pressure of CO_2 in seawater depends on a number of properties including temperature. The warmer sea surface temperatures at the equator and the consequent increased partial pressure (all things being equal) provide some of the explanation of the escape of carbon dioxide in the equatorial zone. However, sea surface temperature alone is not the whole story. Biology also plays a role.

In the sunlit regions of the ocean, phytoplankton combine dissolved inorganic carbon from the water with nutrients to form organic material that cannot participate in the air–sea exchange process. This draw down of inorganic carbon leads to replacement from the atmosphere. The sinking, out of the sunlit zone, of the dead organic material, provides a mechanism by which carbon is removed from the surface ocean.

The flux of CO_2 across the sea surface can be expressed in terms of the difference in density some few metres each side of the interface, $\Delta\rho$. Thus

$$F = C_p U \Delta\rho \tag{5.1}$$

where U is the wind speed (typically measured 10 m above the sea surface) and C_p is a non-dimensional flux coefficient. There remains considerable uncertainty in the value of C_p, and current prescriptions can be found in Wanninkhof (1992) or Liss and Merlivat (1986). The difference in density, $\Delta\rho$, can be related to the partial pressure difference, Δp.

By measuring the wind speed and the carbon dioxide partial pressure, Takahashi *et al.* (2002) could construct a picture such as Figure 5.1 from the observed data. We see in the figure that strong regions of outgassing are concentrated in the equatorial Pacific and the northern Indian Ocean.

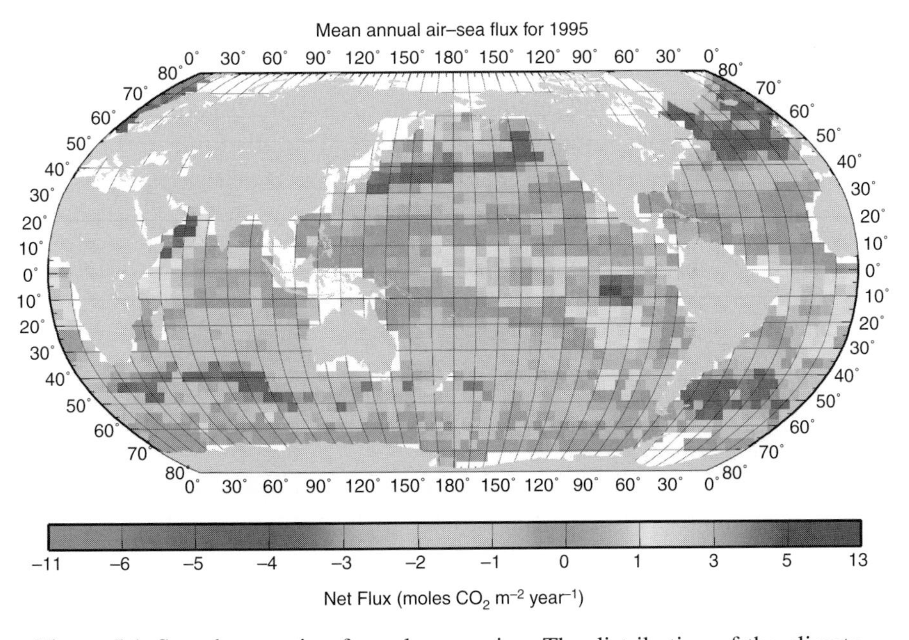

Figure 5.1 See plate section for colour version. The distribution of the climatological mean annual sea to air CO_2 flux in moles of CO_2 per square metre per year. A mole of CO_2 has a mass of 44 g. Reproduced from Takahashi *et al.* (2002).

When the flux of carbon dioxide into the ocean is calculated by averaging over all regions and a number of years, the ocean appears at present to be a sink of about 2 GtC (8 $GtCO_2$) per year.

Ocean circulation

There are two large-scale circulations in the ocean. Each ocean basin has a large horizontal flow or ocean gyre that carries cold, carbon-rich water from the high latitudes to the equator where, as we just discussed, some of the carbon dioxide leaves the ocean. On the western boundaries of the ocean basins, the surface water, warmed by its period in the tropics, flows towards the poles, warming the coastal land mass (Figure 5.2). Variations in the heating by the sun or changes in the surface winds driving these flows, as a result of climate change, will induce many changes in the ocean.

As well as large horizontal flows, there is a smaller but very important vertical circulation. The surface water of the ocean is isolated from the deeper ocean by a region of rapid density change that occurs at depths of 100 metres or so. This strong gradient of density, known as the thermocline, inhibits vertical mixing. The surface water is replaced on a timescale of 10 years or so by water from the deeper ocean, partly by upwelling at certain coastal regions such as Peru or Morocco, and partly by mixing across this density gradient. This vertical flow is known as the thermohaline circulation and provides the source of nutrients for most of the plants in the sea. The replacement time of the deep water is measured in hundreds of years, as its volume is about 50 times that of the surface ocean.

If there were a significant reduction of the vertical circulation or even cessation all together, it might lead to an abrupt climate change. The argument goes that as

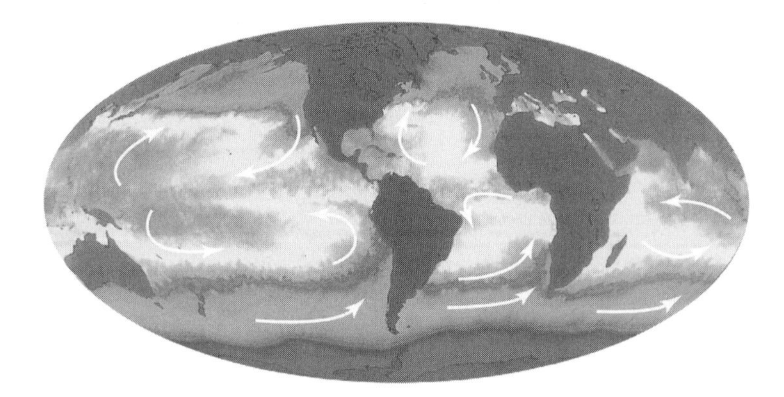

Figure 5.2 See plate section for colour version. Wind-driven surface currents in each ocean basin.

the temperature rises and the density of the surface water near the poles decreases, the water that currently drives the circulation may not be dense enough to sink to the seafloor in the polar regions. This would have significant consequences for humankind, as a major source of protein, that is, fisheries, would be diminished. The winds drive both the upwellings and the surface circulations. Shifts in wind patterns as a result of climate change will alter the surface circulations and, together with the differential heating of the land relative to the sea, will alter the location and strength of upwelling regions. Regional fisheries will feel this environmental change.

We have a picture of a parcel of fluid entering the surface ocean carrying a certain amount of dissolved inorganic carbon. While in the surface ocean this water makes its way around the ocean, taking up or losing carbon to the atmosphere depending on the temperature, salinity and biological activity. After a decade it is downwelled for a journey in the deep ocean that, on the average, takes hundreds of years. It is valuable to recognise this life history since a loss of carbon to the atmosphere in one place will have implications for the water properties elsewhere. We will see that, as well as carbon, the export of the nutrients during the period that the water parcel is in the surface ocean is an important issue in the local carbon flux.

CO_2 partial pressure in the ocean

When the deep water comes to the surface it has a certain partial pressure of carbon dioxide, and this is important in determining the flux of carbon dioxide to or from the atmosphere. The partial pressure of carbon dioxide in the air is rising because emissions of fossil carbon dioxide are greater than the movement of carbon into sinks. The deep water was last in contact with the atmosphere many years ago when the partial pressure of the air was less. Added to the carbon in the water when it left the surface is the carbon from the detritus that has been remineralised while falling through the deep ocean. The remineralisation process uses oxygen, and so oxygen concentration can be used as a measure of the 'age' of deep water. The organic waste from the surface ocean sinks across the thermocline, carrying with it nutrients as well as carbon. This process is known as the biological pump. This export of nutrients leaves most of the surface without significant macronutrients to support further biological activity.

In a global average sense, the recently upwelled water, after the biological pump has removed some carbon, has a lower partial pressure than the air. Because of this the ocean is a sink of carbon dioxide, and this process is known as the solubility pump. As the carbon dioxide partial pressure in the air continues to rise ahead of that of the upwelled water, we expect the solubility sink of the ocean to increase over the next 100 years. If the atmospheric concentration of carbon dioxide were to stabilise, the solubility pump would eventually stop. In the future, deep water will contain more total carbon than at present (because it departed the ocean surface in equilibrium with a post-industrial concentration of CO_2). The changing ocean surface temperature will also influence the yearly ocean solubility sink. A higher temperature than at present will cause the solubility pump to be less effective.

As the concentration of carbon dioxide rises in the surface ocean due to the higher partial pressure in the air, so does the acidity of the water. The first reaction of carbon dioxide in solution is

$$CO_2 + H_2O = H^+ + HCO_3^- \quad (5.2)$$

The right-hand side is carbonic acid, and because of the availability of H^+ ions, it has a low pH. Remember pH is the negative of the logarithm of the concentration of H^+ ions.

The lowering of the pH as a result of increasing atmospheric carbon dioxide concentration will influence those organisms that have shells. The majority of marine calcification occurs in planktonic organisms, and the Royal Society (2005) has recently focused attention on the impact this might have on the future ocean.

Increased ocean sinks

Carbon dioxide waste from fossil fuel burning is presently being dumped in the atmosphere. The mass of air surrounding the earth is 5×10^6 Gt, while the mass of the ocean is 1300×10^6 Gt. The much greater mass of the ocean suggests it will be a better repository of the carbon than the atmosphere.

There are three approaches to increasing the sequestration of carbon in the ocean – alkalinity shifts, direct injection and Ocean Nourishment. The first aims to change the surface ocean chemistry so it can hold more carbon in solution, while the last aims to export carbon to the deep ocean to lower the ocean partial pressure of carbon dioxide at the sea surface and so increase the CO_2 flux from the atmosphere.

Alkalinity shifts

If the partial pressure of carbon dioxide in the ocean surface water were lowered, more carbon would be fluxed from the atmosphere to the ocean. The partial pressure increases with the temperature, and decreases with pH and total alkalinity. This total alkalinity (also called titration alkalinity) is mostly dependent on the concentration of bicarbonate ions present in the seawater. Increase the number of bicarbonate ions without a flux of carbon and you lower the partial pressure

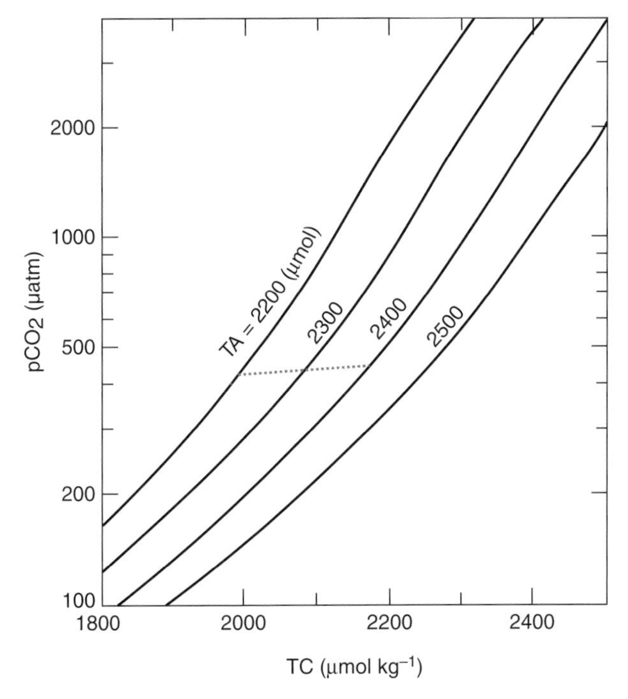

Figure 5.3 The partial pressure of carbon dioxide as a function of total inorganic carbon. Lines of constant pH are almost horizontal with lower pH at higher partial pressures (more acidic). Salinity 35 g kg^{-1}; 15 °C; atmospheric pressure. Adapted from Baes *et al.* (1985).

of carbon dioxide. Removing bicarbonate by converting it to calcium carbonate, say by producing shells for phytoplankton (calcification) lowers the alkalinity and increases the CO_2 partial pressure (Buitenhuis *et al.*, 2001). Increased partial pressure would cause outgassing of CO_2 (i.e. increase the flux into the atmosphere), while lowered partial pressure in the water would increase the flux into the sea.

This can be seen in a plot of the partial pressure of CO_2 in the water as a function of Total Carbon (TC) concentration (Figure 5.3). The carbon is mostly in the bicarbonate ion of Equation (5.2). Total Alkalinity (TA) is measured as the number of moles of alkalinity per kg and can be approximated as

$$TA = [HCO_2^-] + 2[CO_3^{2-}] \tag{5.3}$$

If the water is to store more carbon while staying in equilibrium with the atmosphere, that is increasing Total Carbon, it requires an increase of Total Alkalinity. Very roughly, one mole increase of Total Alkalinity increases the carbon stored by one mole.

Increase the partial pressure of carbon dioxide in the atmosphere and Henry's Law says that the partial pressure in the surface water wants to rise to match the atmospheric pressure. Such a flux of carbon dioxide increases the Total Carbon but makes no change in the Total Alkalinity. Figure 5.3 shows that the pH then falls as the water becomes more acidic. This is the process occurring at present and is the cause of the approximate 2 Gt yr^{-1} of carbon entering into the surface ocean and being subducted into the deep ocean. This increase in acidity that is occurring due to the rising partial pressure of CO_2 will eventually be neutralised by dissolution of carbonate deposits on the seabed. However, these processes are too slow to be useful for managing carbon dioxide in the next few hundred years.

If the carbon were introduced to the ocean in the form of bicarbonate, it would increase both the Total Carbon and the Total Alkalinity. Thus, more carbon can be held in the upper ocean at the same atmospheric partial pressure. Limestone, i.e. $CaCO_3$, could be used as in the equation below:

$$CO_2 + H_2O + CaCO_3 \rightarrow Ca^{2+} + 2HCO_3^- \qquad (5.4)$$

(Magnesium carbonate, $MgCO_3$, is another abundant mineral that can be converted to $Mg(OH)_2$ and then reacted with CO_2. This is considered in Chapter 6, or see Goff and Lackner, 1998.)

Since a mole of carbon dioxide weighs 44 g, and calcium carbonate has a molecular weight of $40 + 12 + 3 \times 16 = 100$, it requires 2.3 tonnes of limestone and 0.4 tonnes of water for each tonne of carbon dioxide sequestered. The Total Carbon and the Total Alkalinity both rise by the same amount.

Caldeira and Rau (2000) suggested that CO_2-rich gas streams (such as flue gases) could be passed through a porous bed of limestone that would be kept wet by water sprays. If the flue gases were from a power station that drew cooling water from the sea, the large amount of water that is needed could readily be supplied. This latter idea has been termed *carbon capture and ocean storage* (CCOS). Rau and Caldeira (1999) estimate that sequestration costs in US dollars of one tonne of carbon dioxide are as follows:

2.3 tonnes of $CaCO_3$ at US$4 per tonne	$9.20
Grinding at 9 kWh t^{-1}	$1.45
Transport (location dependent; say)	$5.00
Extra pumping of cooling water, 57 kWh	$3.99
Capital (neglected)	
Operation and maintenance	$0.50
Total operating cost per tonne CO_2	$20.14

Their analysis produced a range of costs of between US$18 and 128 per tonne of carbon dioxide, depending mostly on transport and water costs. There would be some additional cost penalty in the 78 kWh of energy per tonne CO_2 estimated to be needed for the process, and we can assume that this energy will generate some 78 kg of CO_2. The net sequestration of the storage of one tonne will be 922 kg of carbon dioxide. Transport and construction will also generate some carbon dioxide which would need to be allowed for.

Shifting the alkalinity of the surface ocean is a promising option when there is a concentrated supply of carbon dioxide (such as in flue gases) to make the reaction above occur at useful speeds. When this carbon dioxide is near the seaside and there is already a flow of seawater for cooling, such as in a coastal power station, CCOS is an attractive boutique method of sequestering CO_2. In general the amount of carbon sequestered depends on the volume of seawater flow. Seawater has a calcium carbonate saturation constant of about 4, and so if one wishes to raise the saturation of calcium carbonate in the cooling water by a measured amount, only limited carbon would be sequestered. The positive element of CCOS is that the capital cost would seem to be low for coastal power stations.

For a scheme such as that of Figure 5.4 that uses the existing cooling water channel, one can make an order of magnitude estimate of the capital costs. It does not seem to be large. Design issues are the size of the bubbles produced by flue gas injection into the cooling water, and the diameter of the grains of calcium carbonate. One would like most of the CO_2 to be in solution before the flue gas bubble reached the surface of the cooling water channel. The reaction rate depends primarily upon the pH of the carbonic acid solution and the surface area of the calcium carbonate. If fine ground $CaCO_3$ is used, one would like the grain to be totally dissolved in the turbulent channel flow before the exit to the sea.

When the water with altered chemical properties is returned to the coastal ocean, it will have a higher concentration of carbon, mostly in the form of bicarbonate. As Equation (5.3) shows, both the Total Alkalinity and Total Carbon rise together. If the temperature does not change, the partial pressure will be much the same after neutralisation as when the water was taken from the sea. This can be seen from

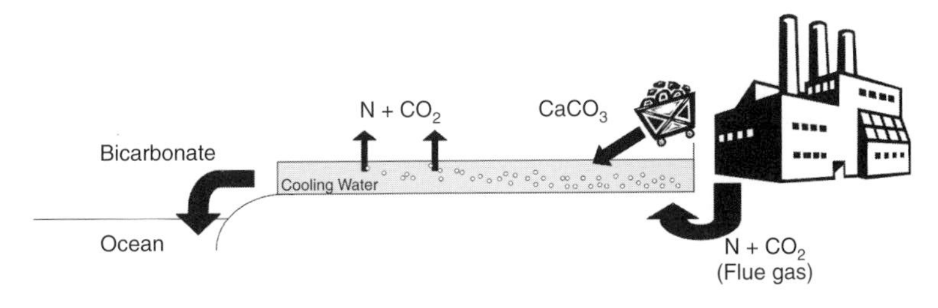

Figure 5.4 The steps in CCOS by alkalinity shifting.

Figure 5.3 from the broken line. However, if cooling water is used, it is returned to the sea at an elevated temperature and care will be needed to calculate the carbon permanently stored.

The environmental impacts of increasing the concentration of calcium carbonate in the CCOS discharge water need to be considered. The saturation state of calcium carbonate, Ω, can be defined as the molar concentration of Ca^{2+} ions multiplied by the molar concentration of carbonate ions [CO_3^{2-}] divided by their product at saturation in pure water. Thus if the Ca^{2+} ions and the carbonate ions are both twice the saturation value, the saturation state $\Omega = 4$. This is a typical value for the surface water of the world's oceans.

Adding calcium ions to the ocean surface water will increase Ω. There is a limit to the amount of calcium bicarbonate that can be kept in solution in seawater, but simple experiments show that saturations of at least $\Omega = 16$ do not lead to spontaneous precipitation. The other ions in seawater are believed to be the reason for stable supersaturation (with respect to freshwater). With time, the cooling water plume with the elevated Ω will mix with the ocean water and the value of Ω will decline.

Calcium carbonate shells are an important element of many marine organisms, and the rate at which they can carry out calcification depends upon the saturation state Ω (Delille *et al.*, 2005). This seems to have first been demonstrated for net community calcification rate by Langdon *et al.* (2000) using the BIOSPHERE-2 facility operated by Columbia University. The influence of calcium carbonate saturation on calcification rate can be seen in Figure 5.5. As can be seen from Figure 5.3, the pH

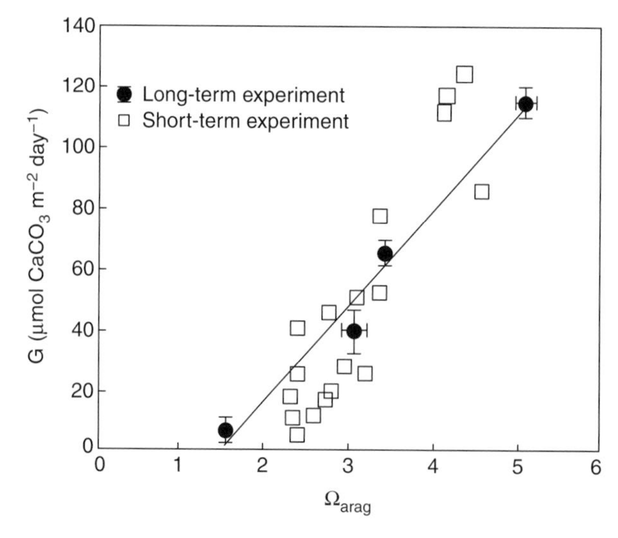

Figure 5.5 Effect of calcium carbonate saturation state on the calcification rate, G, of coral reef ecosystems. Reproduced from Langdon *et al.* (2000).

will fall with a flux of carbon into the sea without neutralisation. This has been shown by Marubini and Thake (1999), for example, to have a deleterious effect on calcification. A higher concentration of bicarbonate will encourage the growth of shellfish (although there is a level where there is already ample bicarbonate and other factors limit growth). If the shellfish prosper in the stream of enriched seawater, they will be converting the bicarbonate back to calcium carbonate and carbon dioxide. This loss of net sequestration can often be neglected, but makes us realise that the growth of coral is a source of greenhouse gas.

For carbon credits generated by CCOS with neutralisation to be traded freely, this method of sequestering carbon needs to be approved by the UNFCCC. The Executive Board of the Clean Development Mechanism (CDM) of the UNFCCC currently (2008) has CCOS under consideration.

Direct injection into the ocean

Carbon dioxide can be captured and directly injected into the ocean at depth. Injection into the surface waters is of little value, as the result is a rise in the partial pressure of carbon dioxide in the water, which leads to degassing. Injected into the deep ocean, carbon dioxide must await the return of the surrounding water to the ocean surface before it can communicate with the atmosphere. The sink capacity of the ocean has been estimated as between 1400 and 20 000 Gt of carbon by Haugen and Eide (1996), the same order as the expected fossil-fuel reserves, estimated by Hoffert *et al.* (1979) at about 7000 Gt.

Hertzog (1999) has mentioned a proposed *proof of concept* demonstration of direct injection of carbon dioxide to reduce the technical uncertainties, and this was under investigation by an international group under the auspices of the International Energy Agency Greenhouse Gas (IEAGHG) research and development programme. Permits were not obtained for these experimental injections of carbon dioxide. Direct injection relies on the fact that carbon dioxide is a liquid more dense than water at the temperatures and pressures of the deep ocean. Since the vertical circulation of the ocean returns deep water to the surface where it can again equilibrate with the atmosphere, direct injection is a temporary escape from the dangers of high levels of carbon dioxide in the atmosphere.

So far, several methods have been proposed for the direct injection of CO_2 into the deep ocean, such as using dry ice or a dense downhill plume (Marchetti, 1977; Adams *et al.*, 1995; and Kosugi *et al.*, 1999). After considering cost performance, the most promising ocean CO_2 sequestration can currently be categorised into two methods, namely mid-depth dissolution (Ozaki, 1997) and deep storage (Shindo *et al.*, 1995; Brewer *et al.*, 1997; and Aya *et al.*, 1999). These are shown in Figure 5.6.

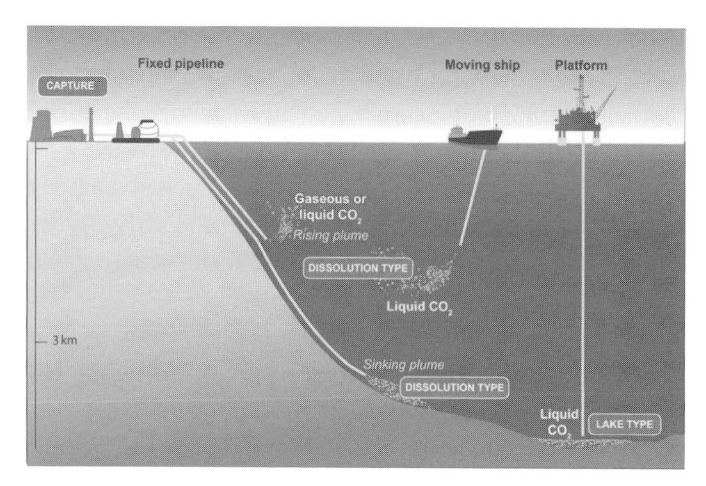

Figure 5.6 See plate section for colour version. Schematic scenarios in direct injection. From Metz *et al.* (2005).

It is said that the current technology is able to cope with mid-depth injection without important technical innovations. The CO_2 sequestration scenario proposed by Nakashiki *et al.* (1995) and Ozaki (1997) is that CO_2 is captured and liquefied at a few 100 MW-class power plants, shipped by liquid CO_2 tankers to an offshore station, transferred to other vessels which are equipped with a long vertical pipe, and injected into the deep ocean at the rate of 0.2 tCO_2 sec^{-1} (6 million tonnes per year).

The necessary energy, cost and efficiency of the middle-depth dissolution method is estimated by the Research Institute of Innovative Technology for the Earth to be about 400 kWh per tonne of CO_2, US$65 per tonne of CO_2, and 87% respectively. The efficiency means that 130 $kgCO_2$ is emitted when 1 tonne of CO_2 is disposed of, based on the (low) CO_2 emission intensity, 0.37 $kgCO_2$ kWh^{-1}. The assumption includes the usage of liquefied natural gas (LNG) power plant, monoethanol absorption of CO_2, and liquid CO_2 tanker shipping to 500 km offshore from the plant, as is shown in Figure 5.7.

There are two more issues we need to settle, i.e. sequestration time and biological impact. The globally averaged vertical rise-speed of a chemical species in the deep ocean is a few metres per year. (This is consistent with the surface waters being replaced on the timescale of a decade.) The sequestration time, in other words the residence time of CO_2, can be calculated simply by using this, although there must be regional variations. Long-residence-time sites are pointed out by Ametistova *et al.* (2002), in particular warm western currents, e.g. Gulf Stream and Kuroshio, and deepwater formation regions, e.g. North Atlantic and Southern Ocean. Ocean circulation models are able to elucidate the sequestration time of each injection site. For injection at 1500 m at some sites (like NewYork), half the

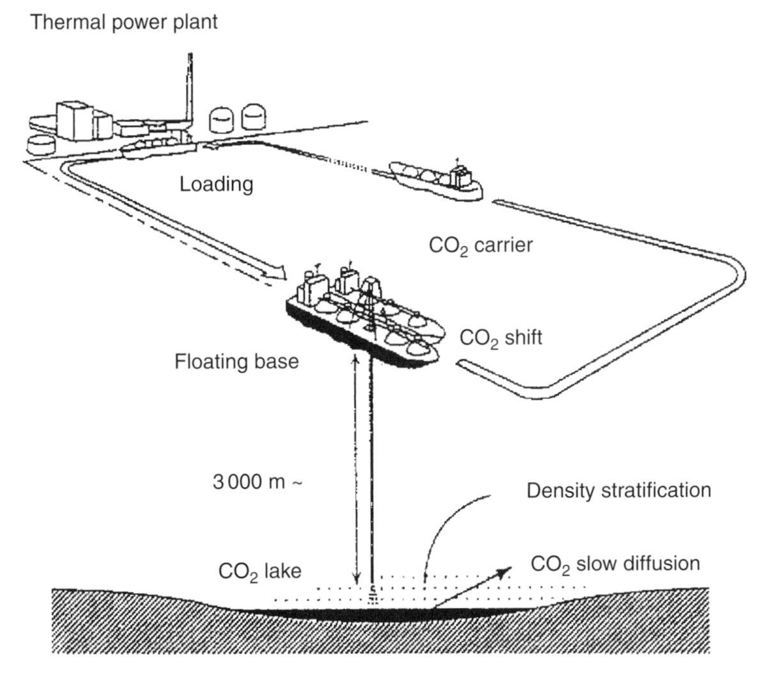

Figure 5.7 A conceptual diagram of storing CO_2 in a lake at the seafloor, after Ormerod, 1997.

carbon has returned in 200 years, but for most of the sites it takes more than 500 years (Orr and Aumont, 1999). In the case of the deep storage on the seafloor, it is believed that the sequestration time is very long, because a CO_2 hydrate film is supposed to cover the interface between liquid CO_2 and seawater and to slow the dissolution rate of CO_2. Moreover, recent measurements done by Hirai *et al.* (1997) and Aya *et al.* (1997) show that there is a moderate rate of CO_2 dissolution through the hydrate film. Carbon dioxide hydrate is a crystalline molecular complex involving CO_2 and water.

In the case that we have a lake of carbon dioxide, denser than the surrounding water, Thornton and Shirayama (2001) have considered the impact of CO_2 on benthic organisms beneath the lake. The deep storage of CO_2 extinguishes marine life in the immediate vicinity of CO_2 lakes. Brewer (2000), in a seabed experimental injection of carbon dioxide, found fish swimming near the injected carbon dioxide without obvious ill effect. The major impact may be confined to a manageable region of the seafloor. One early examination of the impacts on the atmospheric concentration of direct injection is provided by Hoffert *et al.* (1979).

Next we focus on the concept of dissolution of carbon dioxide at mid-depths, because it is believed that this method provides the marine ecosystem with the least impact amongst the above-mentioned direct injection options. One of the uncertainties in this method is its local impact on marine organisms in the vicinity of the injection points. To mitigate this uncertainty, wide dilution is desirable. In the deep ocean, where the average current speed is said to be less than a couple of centimetres per second, we cannot expect dilution to be as rapid as in the upper ocean. Accordingly, it is important to investigate the dilution process of dissolved CO_2 over distances of hundreds to thousands of metres.

Sato *et al.* (2002) carried out numerical simulations of liquid CO_2 dissolution and dissolved CO_2 dilution by using a computational fluid dynamics model. Figure 5.8 denotes the contour maps of dissolved CO_2 concentration at 1, 2, 5, and 10 hours after the start of injection. The injection depth is set to be 2000 m. The density of liquid CO_2 is lighter than that of seawater at a depth less than 3000 m.

At a depth of around 2000 m, ejected liquid CO_2 forms a rising droplet plume entraining surrounding seawater. During the rise, CO_2 dissolves into the entrained seawater and the resulting denser water peels out from the plume and sinks. The coexistence of the upward droplet plume and the downward dense solution results, above the injection point, in an unsteady flow structure like a fountain. The dense water continues to sink to the depth where the density of the stratified ocean is equal to the CO_2-enriched water. Then the dense CO_2 solution stops sinking and spreads horizontally as a gravity current.

Sato and Sato (2002) made a prediction of the biological impact, making use of the iso-mortality curve proposed by Auerbach *et al.* (1997). The curve was drawn by accumulating LC50 (lethal concentration, at which 50% of organisms die) and LC90 (concentration at which 90% of organisms die) data with respect to multiple sets of exposure duration to acidification, for marine organisms. The LC50 concept is widely used for testing acute toxicity of water. They found that organisms at the carbon dioxide injection point experienced a minimum pH of about 5, and as they drift with the flow it gradually increases to 7. Also predicted in their simulation was that these marine organisms are on the safe side, with exposure beneath the 50% mortality limit.

For some industrial processes, carbon dioxide is already available as a waste product in concentrated form. This is ideal for injection into the sea as it avoids capture costs. The sequestration time is less in the ocean than for the injection into geological structures. However, for many geotechnically active countries with access to the sea, direct injection into the ocean may be an attractive option. The leakage of carbon dioxide back to the atmosphere comes about when the carbon-dioxide-enriched water from depth returns to the surface.

The leakage makes the calculation of the climate benefit complicated. Placing the carbon dioxide in the water has a cost and has the consequence that the carbon dioxide concentration in the atmosphere is less than it would have been. Less climate change

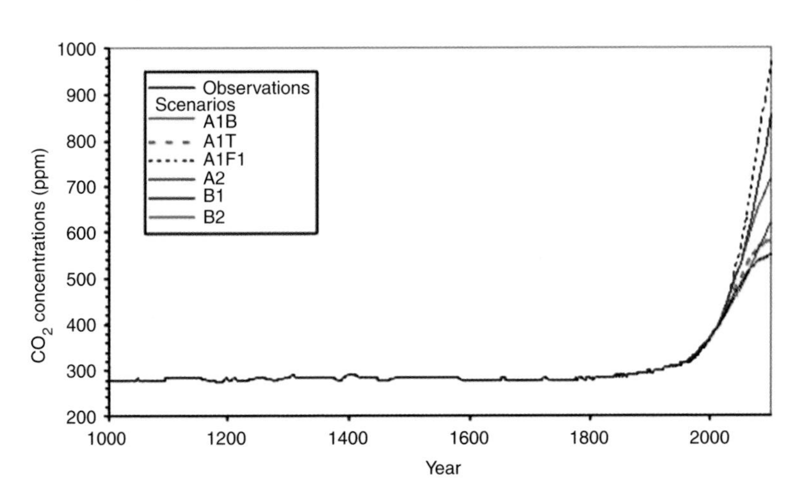

Figure 1.2 The change in atmospheric concentration of CO_2 over the last thousand years. Various future scenarios are also shown. The A2 scenario is described in more detail in the text; for others refer to Nakicenovic and Swart (2000). Note the truncated vertical axis. (Is this a good engineering practice or is it a technique to raise alarm?) Reproduced from Houghton *et al.* (2001).

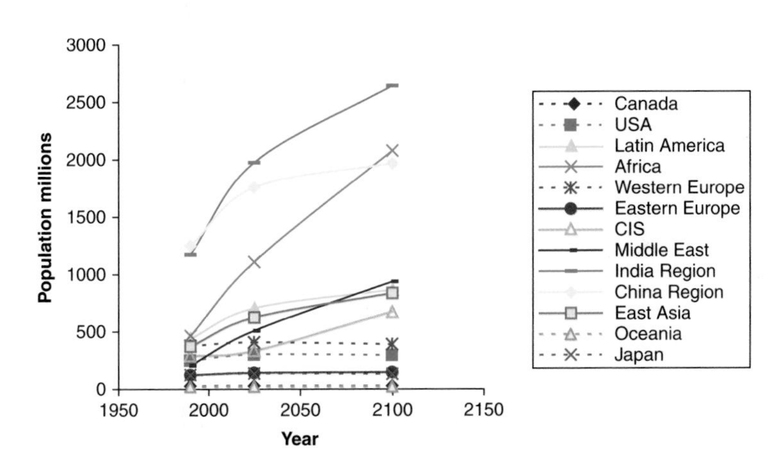

Figure 1.4 Population prediction that Alcamo *et al.* (1994b) used in IMAGE2 for the regions shown in Figure 1.3.

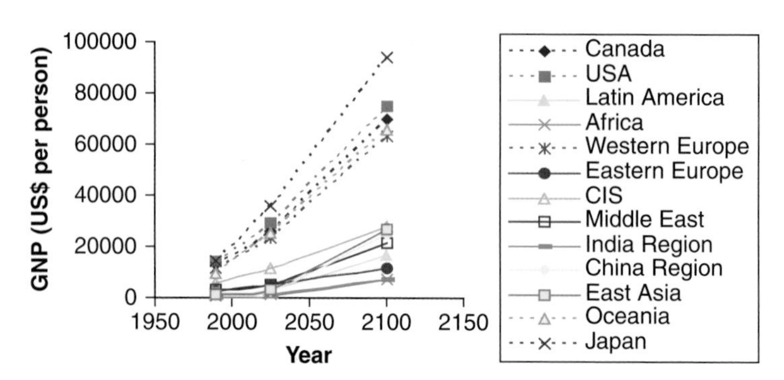

Figure 1.5 Gross national product per person in 13 regions. By the year 2100, Japanese residents are predicted to have an average income of US$95 000. After Alcamo *et al.* (1994b).

Figure 3.1 A wind farm. Image © Pedro Salaverría / www.shutterstock.com.

Figure 3.2 Solar chimney with its solar energy collecting roof made of green material.

Figure 4.3 An artist's concept of the ship to provide seawater droplets to form low-level marine clouds. Rate of water injection 1 kg sec^{-1}. Image John MacNeill.

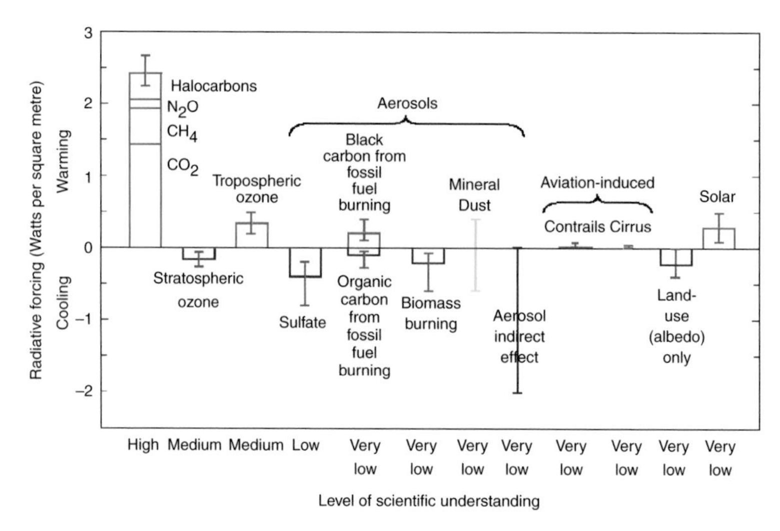

Figure 4.4 The global mean radiative forcing of the climate system for the year 2000, relative to 1750, reproduced from Houghton *et al.* (2001).

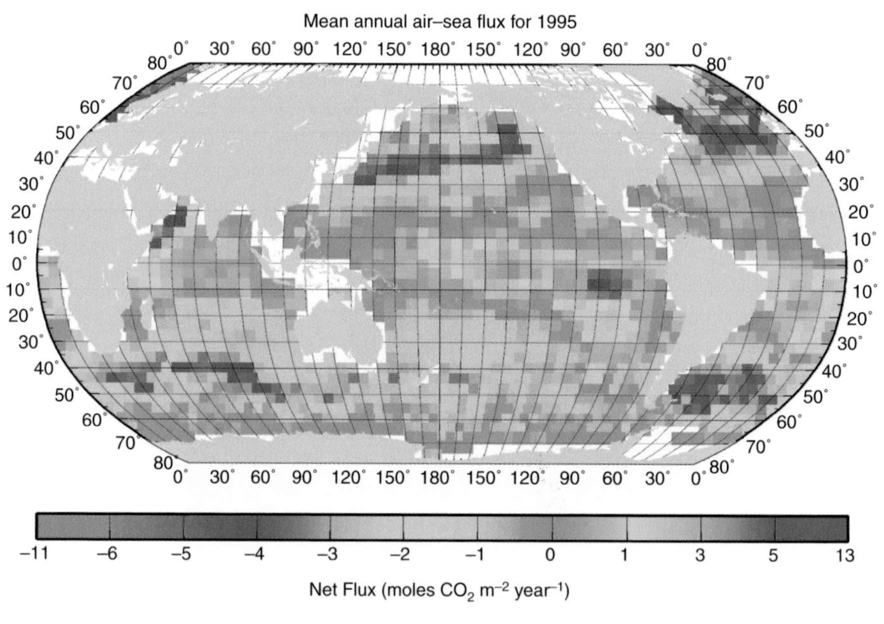

Figure 5.1 The distribution of the climatological mean annual sea to air CO_2 flux in moles of CO_2 per square metre per year. A mole of CO_2 has a mass of 44 g. Reproduced from Takahashi *et al.* (2002).

Figure 5.2 Wind-driven surface currents in each ocean basin.

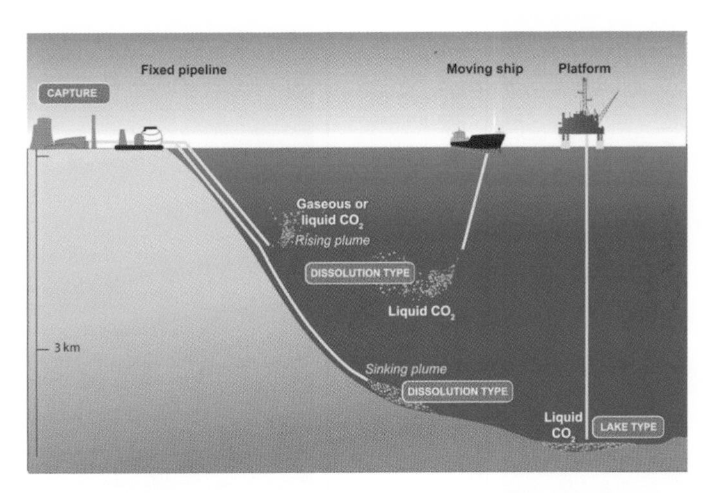

Figure 5.6 Schematic scenarios in direct injection. From Metz *et al.* (2005).

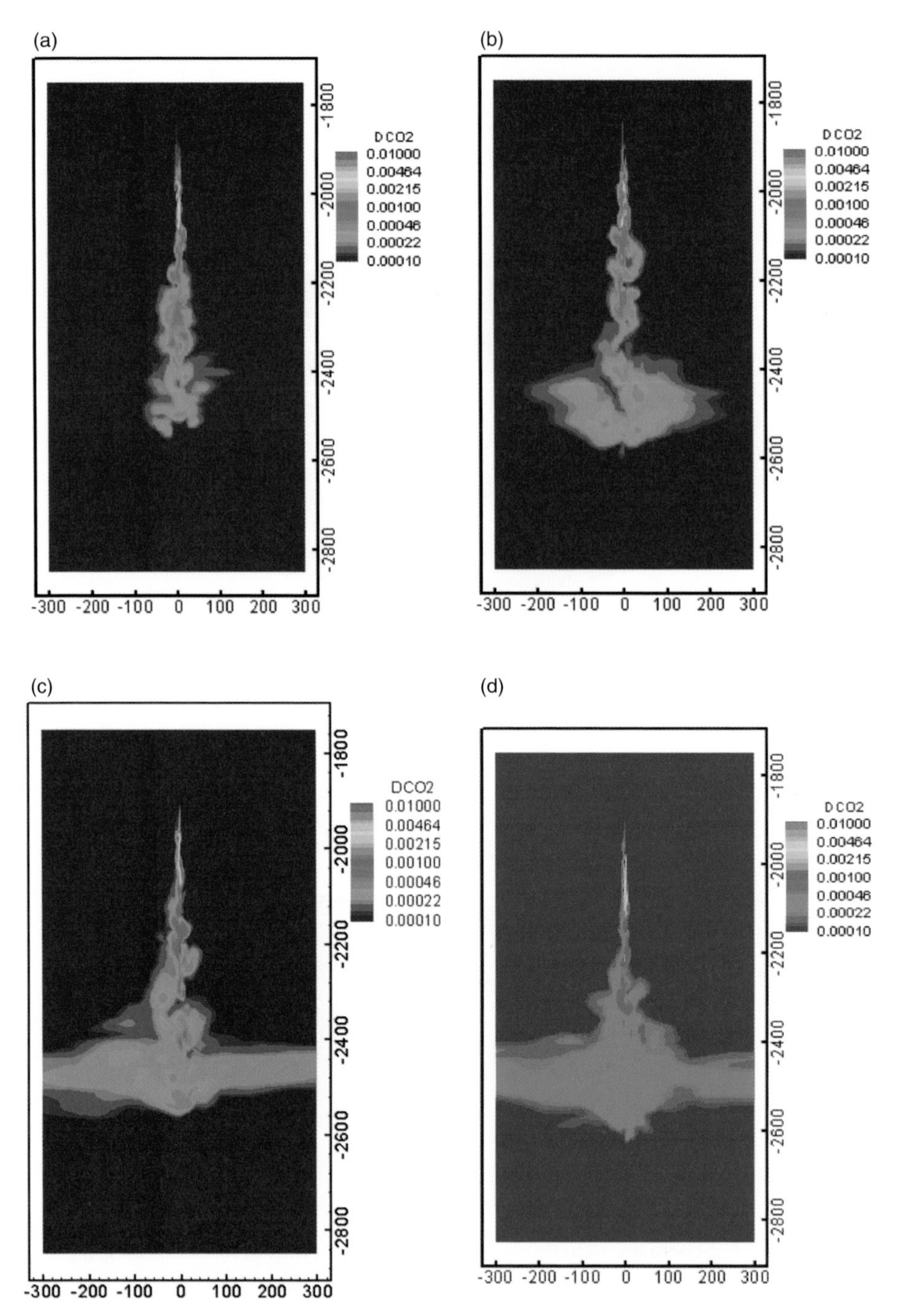

Figure 5.8 Contour maps of dissolved CO_2 concentration (DCO2) at (a) 1; (b) 2; (c) 5; and (d) 10 hours after start of injection. Distance is in metres. Injection at a depth of 2000 m.

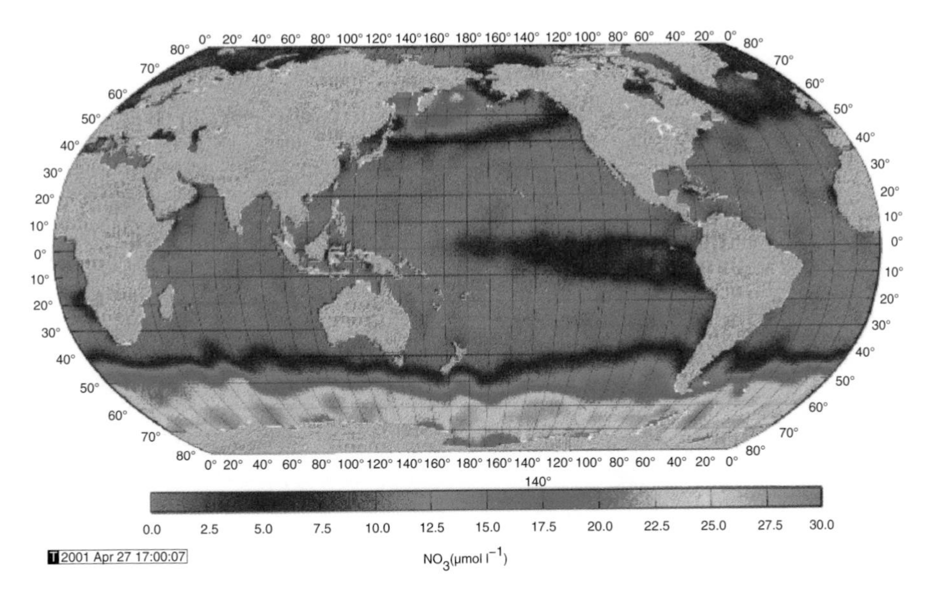

Figure 5.11 Average surface nitrate levels in the upper ocean, after Levitus *et al.* 1994.

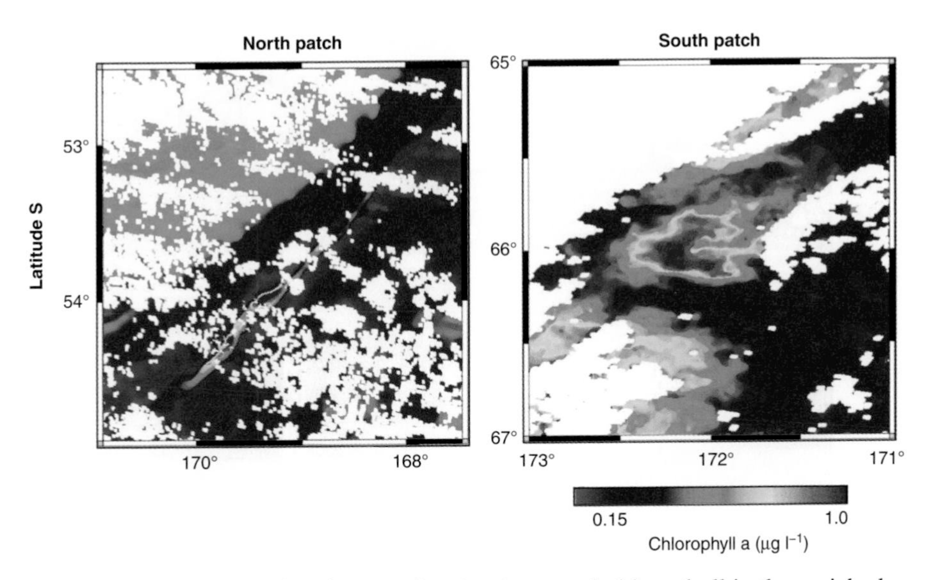

Figure 5.13 Ocean colour images showing increased chlorophyll in the enriched patches as red, after Coale *et al.* (2004).

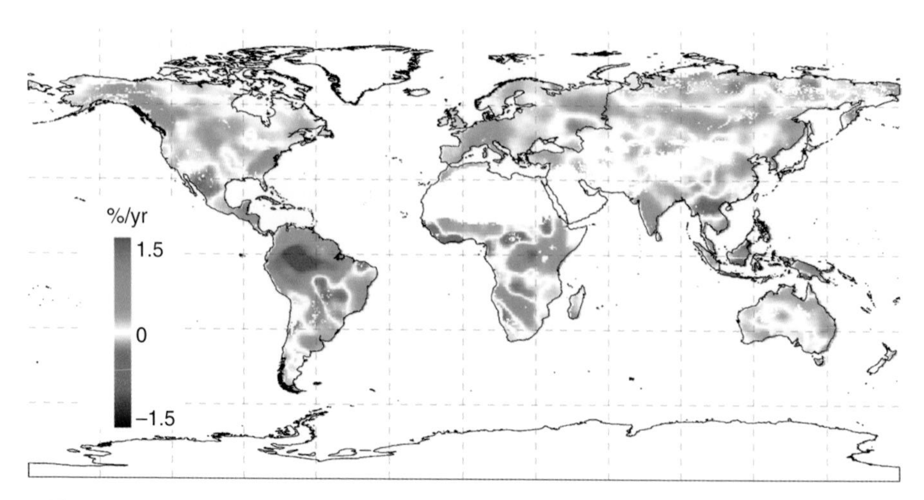

Figure 6.1 The changes of net primary production calculated from a productivity model, after Nemani *et al.* (2003).

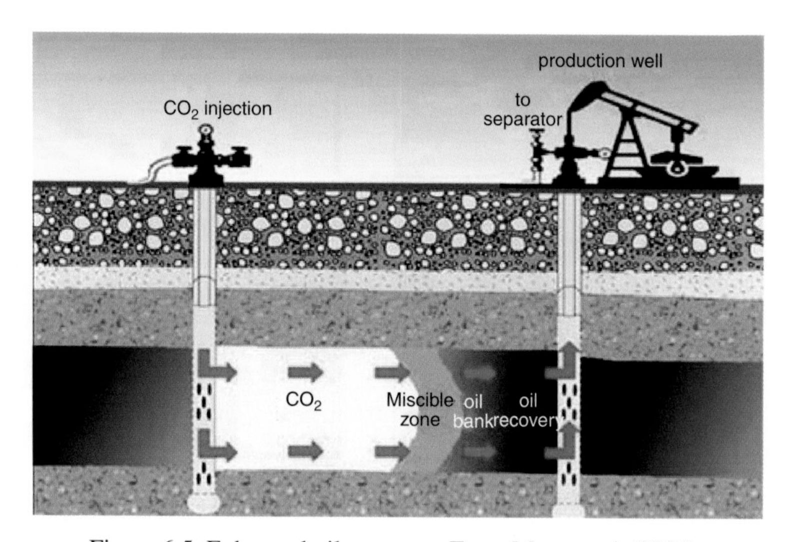

Figure 6.5 Enhanced oil recovery. From Metz *et al.* (2005).

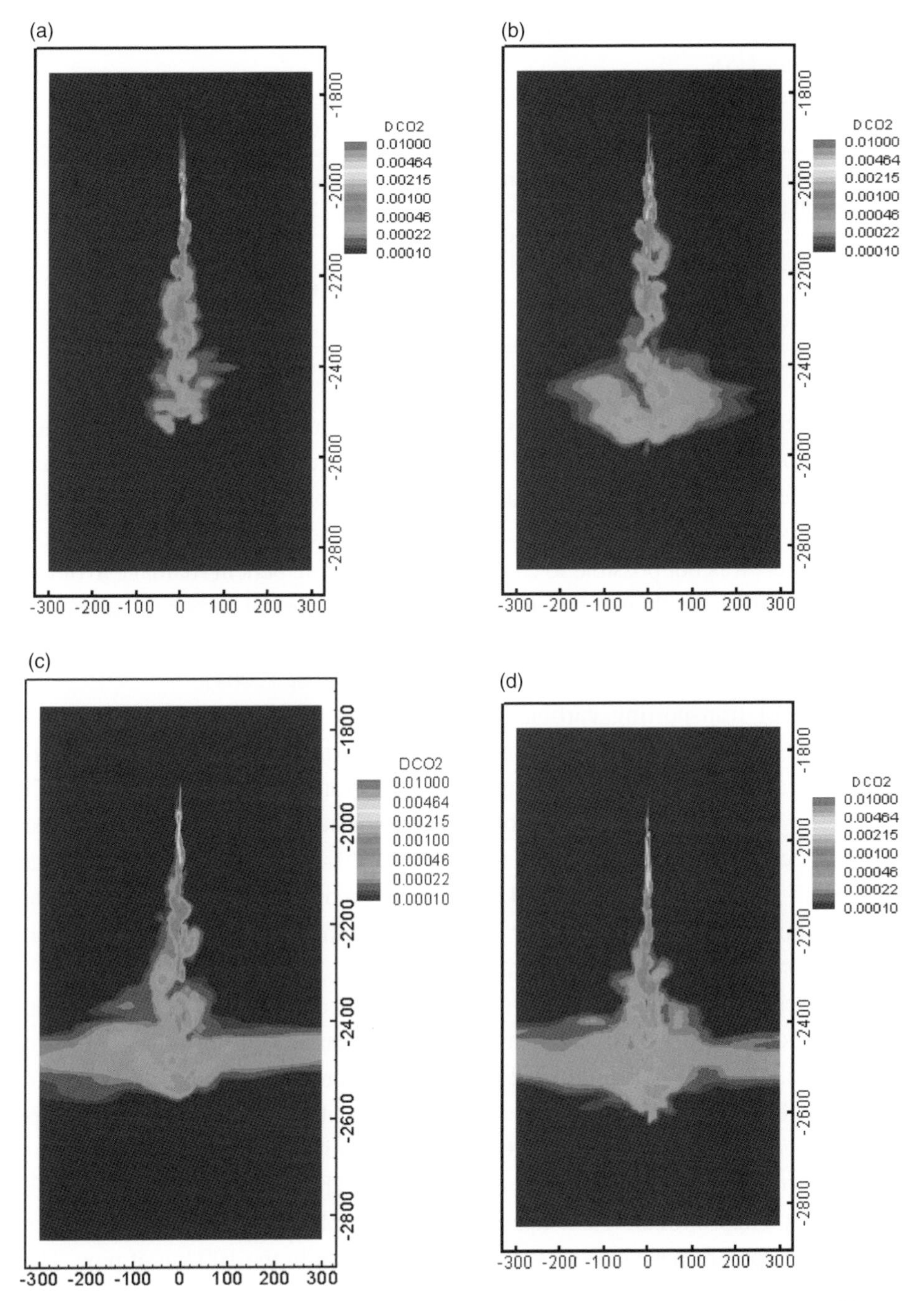

Figure 5.8 See plate section for colour version. Contour maps of dissolved CO_2 concentration (DCO2) at (a) 1; (b) 2; (c) 5; and (d) 10 hours after start of injection. Distance is in metres. Injection at a depth of 2000 m.

Table 5.1. *Ocean storage of captured carbon dioxide – transportation costs (only) in US$.*

			Cost per tonne CO_2
Sequestration	18 million	t yr^{-1}	
Cost of capital	$12 million	$ yr^{-1}	$0.66
Pumping cost	$4 million	$ yr^{-1}	
O&M	$0.5 million	$ yr^{-1}	
Total	$16.5	$ yr^{-1}	**$0.9**
Carbon dioxide generated	40 000	t yr^{-1}	

is to be expected and we will discuss this through the proxy of the surface temperature. The economic damage caused by climate change is considered in Chapter 7, and so we can discuss the damage avoided by direct injection of carbon dioxide into the ocean. Since the economic damage is a complicated function of the concentration of carbon dioxide, it is not possible to calculate the economic benefit (damage avoided) without assuming a baseline or *business as usual* time history of carbon dioxide in the atmosphere. An interesting effort to assess the value of direct injection of a tonne of CO_2 into the ocean has been undertaken by Tokimatsu *et al.* (2004).

The cost of transporting carbon dioxide from a coastal site to an injection point 500 km offshore could be estimated from considering the cost of two ships (US$50M each) capable of providing an average rate of delivery of 7000 tCO_2 day^{-1}. We will use an interest rate of 10% and assume 350 days operation per year. This cost of the capital is about $(10\% \times 100 \times 10^6)/(7000 \times 350) =$ US$4 per tonne. Such vessels cost some US$14 000 per day to operate (US$2 per tonne) and generate about 50 tonnes of CO_2 per day from burning 15 t d^{-1} of fuel. There are other costs such as refrigeration and injection.

Another estimate of the costs of direct disposal in the ocean is to consider a pipeline from shore to mid-depths. Pipe laying has been carried out in waters of greater than 2000 m. The cost of a mild steel pipe is of the order of US$1 million per km laid and, with suitable injection nozzles and pumps, a site that needed 100 km of pipe might involve a capital cost of US$120 million. Servicing this capital at 10% pa is a yearly cost of US$12 million, as shown in Table 5.1. The weight of the carbon dioxide in the pipe approximately balances the 300 atmosphere pressure at the injection nozzle, so the pumps have only to overcome the friction in the 100 km of pipe. This is a function of the speed of the carbon dioxide within the pipe. A reasonable energy consumption is 40 million kWh at a cost of US$4M and a flow of 18 Mt yr^{-1}. This is US$1 per tonne of CO_2. Certification and monitoring costs need to be considered. To this must be added the capture costs and an allowance for the carbon dioxide generated by manufacturing the pipe and providing the fossil fuel for the pumps.

The above simple examples suggest that the cost of injection is modest compared with the present cost of capturing carbon dioxide from flue gases. Capturing carbon dioxide from the atmosphere at concentrations of order 400 parts per million is even more expensive because of the low concentration; see Chapter 3.

An interesting way to overcome some of the difficulties with direct injection of liquid carbon dioxide has been suggested by Wadsley (1995) and by Golomb and Angelopoulos (2001). They suggest mixing ground-up limestone ($CaCO_3$) with the liquid CO_2 to change the alkalinity of the water surrounding the injection point. The emulsion would have a density greater than seawater and so would sink from the point of release. However, the most important benefit might be that the carbon dioxide and calcium carbonate will form bicarbonate and change the Total Alkalinity in a manner that lowers the partial pressure of the carbon dioxide. This can then be considered permanent sequestration. This idea is the same as the alkalinity shift idea above except that it relies on carbon dioxide capture. The penalty of forming the emulsion is the additional cost of providing the pulverised limestone.

Efforts to carry out proof-of-concept testing of ocean direct injection have been hindered by the intervention of environmental groups who have used legal and political processes to block tests in the ocean.

Ocean Nourishment

Ocean Nourishment is the general term to describe the purposeful introduction of nutrients into the ocean photic zone to increase photosynthesis. Photosynthesis in the ocean is the process of converting the inorganic chemicals in the sea to organic matter. The energy to drive this process comes from the sun, but uses less than $0.1\,W\,m^{-2}$, that is, less than 0.03% of the average solar radiation. This organic matter is mostly carbon, and so increased photosynthesis draws down the inorganic carbon content in the upper ocean. The production of vegetable matter, termed primary production, is the base of the food chain and so regulates the production of other elements in the marine food chain, such as fish.

Martin *et al.* (1990) pointed out that iron limited the export of carbon in the Southern Ocean, and suggested that climate intervention was possible by the use of iron. The concept of nourishing the ocean to both draw carbon dioxide out of the atmosphere and to increase the marine protein was explored by Jones and Young (1997). They considered areas where macronutrients limited primary production.

Both micronutrient and macronutrient nourishment has been carried out on the high seas. The London Convention, which is concerned with dumping waste in the sea, does not prohibit the introduction of substances for purposes other than mere disposal. They have recently adopted an assessment framework for scientific research involving ocean fertilisation. For nourishment in the Exclusive

Economic Zones (EEZs) of sovereign countries, the UN Law of the Sea would seem applicable.

Before we can examine if all the added nutrients can be expected to be exported with the carbon over timescales of a year, we will need to understand the oceanic carbon cycle. For how long is the exported carbon sequestered in the ocean? This and other questions to do with Ocean Nourishment will be addressed in the rest of this chapter.

Changing the ocean primary production

The sunlit or photic zone of the ocean contains about 0.5 Gt of carbon (Falkowski and Raven, 1997), in contrast with the atmosphere which contains about 750 GtC. When there are nutrients present, phytoplankton subdivide and die within two to six days on average. Some of this dead organic matter falls into the deep ocean as particulate organic carbon (POC). Some of the phytoplankton are grazed by zooplankton, and the carbon excreted falls to the deep ocean. The remainder is remineralised in the surface ocean.

The process is illustrated in Figure 5.9.

Consider the left-hand box in Figure 5.9, where new primary production is represented. The sunlight provides the energy for nutrients to be combined with carbon

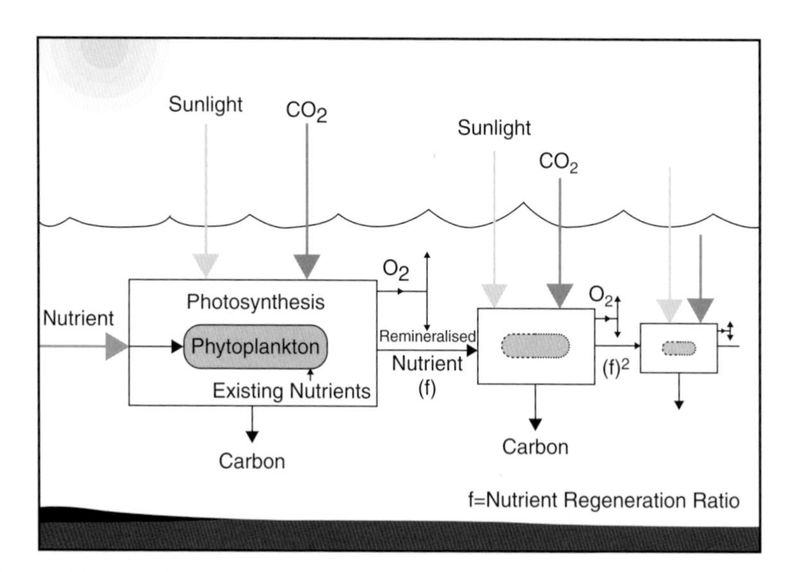

Figure 5.9 The diagram shows how the nutrients lead to the export of carbon from the upper ocean. An alternative pathway through zooplankton is not shown. Eventually all the nutrient that enters the photic zone is exported. The nutrients can be upwelled as nitrate or can be introduced anthropogenically.

to form organic matter. Chlorophyll is essential in this process, and the chemical reaction releases oxygen. There are four pathways for the organic matter. Some of the organic matter is remineralised in the surface photic zone into reactive nitrogen, mostly ammonia, and other constituents including inorganic carbon. These components are again converted to organic matter, in the second box in Figure 5.9. Some is soluble and stays in the surface ocean until subducted or broken down by bacteria. The remainder is either exported as particulate organic matter by sinking under the influence of gravity or is consumed (grazed) by predators such as zooplankton (not shown in Figure 5.9). These animals excrete some of the carbon as faecal pellets. Zooplankton and fish process some of the nitrogen to form urea.

The nutrients, now in the second box in Figure 5.9, undergo the same process again unless there is a limitation of a micronutrient such as iron. With ample iron deposited from the atmosphere, the process continues until all the nutrient introduced on the left of Figure 5.9 is exported out of the photic zone. We know all the new primary production is exported (or subducted) from the surface ocean, as we do not see a build-up of macronutrients such as phosphate despite a continuous supply from upwelling each year.

There is a remarkable consistency in the ratio of chemical elements in marine organic matter. This Redfield ratio of carbon atoms to nitrogen atoms is 6.6 to 1 for typical phytoplankton. This can be understood from an idealised representation of marine photosynthesis in terms of macronutrients (Redfield *et al.*, 1963):

$$
\begin{aligned}
&106CO_2 + 16NO_3^- + H_2PO_4^- + 17H^+ + 122H_2O \\
&\Leftrightarrow (CH_2O)^{106}(NH_3)^{16}(H_3PO_4) + 138O_2
\end{aligned}
\tag{5.5}
$$

When the organic material is formed, oxygen is produced; when remineralisation by bacteria reverses the equation, oxygen is consumed. During dark periods the process reverses, releasing carbon dioxide. When nitrate is involved, hydrogen ions are consumed, reducing the acidity of the seawater.

In the above equation, 16 molecules of nitrate combine with 106 molecules of carbon dioxide. If we add nitrate and phosphate from outside the surface ocean, the nitrogen (atomic weight 14) then sequesters 20.8 units by weight of carbon dioxide (atomic weight 44) for one unit of nitrogen taken up by the phytoplankton.

The deep ocean cycling can be seen from Figure 5.10. It takes an order of magnitude longer than the process in Figure 5.9. As discussed, nutrients new to the photic zone (labelled N) combine with carbon dioxide to form organic matter which sinks into the deep ocean. The deeper it sinks before being converted into soluble inorganic materials, the longer it takes to return to the surface. The returning waters are rich in nutrients because of the constant rain of ‘marine snow’ from the photic zone. A little of the marine snow reaches the seafloor and is sequestered

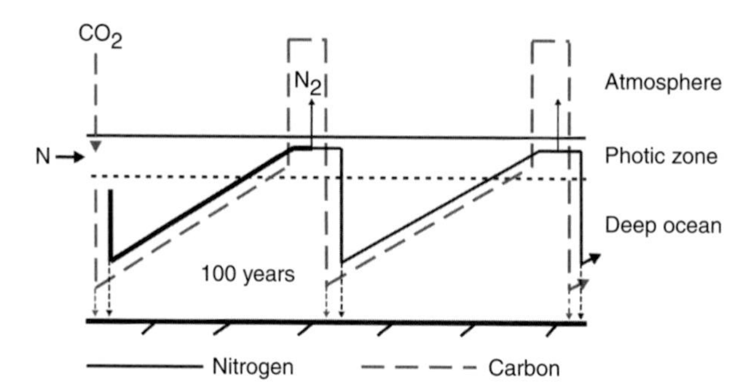

Figure 5.10 The path of introduced nitrogen as it sinks as organic material and is upwelled to the sea surface to be cycled again after a period of some hundreds of years. Time increases to the right. N = nutrients.

there forever, taking with it much of the exported iron which it scavenges on the way. Some of the nitrogen is converted to N_2 or nitrous oxide, N_2O, during this deep ocean circuit. These gases are volatile and readily escape to the atmosphere after reaching the sea surface. The nitrogen lost is believed to be made up in part by cyanobacteria which are able to utilise gaseous nitrogen.

The carbon dioxide initially incorporated into the organic matter is also free to return to the atmosphere, but the nitrate is not. Thus, even though the carbon dioxide pumped out of the surface layer some hundreds of years ago as a result of new primary production may return to the atmosphere, a significant amount of carbon dioxide is again captured by photosynthesis and sent on another circuit through the deep ocean. When iron is lost on the journey from the surface ocean through the deep ocean, the upwelled nutrients cannot always support phytoplankton growth. Iron is replaced in part by precipitation from the atmosphere in the form of dust, but in some regions the surface water is subducted before there is enough iron to allow conversion of carbon dioxide to organic matter. When the formerly organic carbon returns to the atmosphere, it is not recaptured in these high-nutrient, low-chlorophyll regions.

Najjar and Keeling (2000) calculated the flux of oxygen from the present-day new primary production to be approximately 5×10^{14} moles O_2. Remineralisation in the thermocline consumes oxygen (see Equation (5.5)), and when these low-oxygen waters return to the surface, the oxygen consumed hundreds of years before will be replaced by a flux back into the surface ocean. The drop in oxygen of the deeper waters increases with the 'age' of the water. Some nitrogen is converted to N_2O in the remineralisation process, and the amount is believed to be related to the level of oxygen present.

It has long been known, due to experiments by Thomas (1970), that the addition of a broad range of nutrients could induce phytoplankton growth. Tranter and

Newell (1963) showed that the addition of iron could increase the natural assemblage of phytoplankton in the Indian Ocean. Thomas (1969), in the tropical Pacific, and Edington (2005) in the Tasman Sea showed, by incubating in natural light on the deck of a ship in culture bottles, that in some parts of the ocean the addition of the macronutrients nitrogen and phosphorous was enough to cause more primary production. The micronutrients, also needed for plant growth, were in sufficient concentration to allow the growth of more organic matter.

Next we examine the concepts of ocean fertilisation, the process of adding additional nutrients to the upper ocean to increase photosynthesis. There are five ways to provide additional nutrients to the photic zone on a practically important scale. Nutrients are available at the top of the thermocline, iron is readily available from the land, cyanobacteria can be used to fix nitrogen or nitrogen can be manufactured by the Haber–Bosch process. Phosphorus can be mined or found in the surface ocean.

Ocean Nourishment: induced upwelling

The waters near the top of the thermocline, but below the photic zone, are a dilute source of macronutrients. Artificial upwelling of this thermocline water has been considered as a means of fertilising the photic zone. How much water would we need to move, to bring up a million tonnes of nitrogen? We know from measurements that the average concentration of nitrogen in the deep ocean is about 21 μM (21×14 mg m^{-3}), and so a volume of $10^6/294 \times 10^{-9} = 3.4 \times 10^{12}\,m^3$ is required. Using the Redfield ratio, we find that this one million tonnes of nitrogen would support $6.63 \times 44/14 = 20.8$ million tonnes of carbon dioxide capture.

However, with the million tonnes of nitrogen comes a large amount of carbon. Some of the carbon is a result of the solubility pump and some the biological pump. To the first order, the nitrogen is in the Redfield ratio to the carbon exported by biological pump, and all the nitrogen is needed to recapture this carbon by photosynthesis. We will assume the remaining carbon was in equilibrium with the atmosphere when it was subducted many years ago. Let us consider the situation where the atmospheric partial pressure is 400 ppm. If we assume that when the water to be pumped was subducted, the partial pressure was 280 ppm by volume, then there is a difference of 120 ppm when the water is again in the surface ocean. The amount of extra carbon stored in the deep ocean can be estimated. We see from Figure 5.3 that, at a constant Total Alkalinity of 2350 μmol l^{-1}, the difference in partial pressure corresponds to about 70 μmol l^{-1} ($70 \times 12 \times 10^{-9}$ tC m^{-3}). To store 1 t of CO_2 ($12/44 = 0.27$ tC), one needs to pump up $0.27/840 \times 10^{-9}$ m^3 of water. That is, 0.3 million cubic metres of seawater.

To raise the water takes energy. Let us assume we raise water from the thermocline where the water has a density 1.0 kg m^{-3} above that of the mixed layer. If we move this to the centre of the mixed layer over a distance of 100 m, the potential energy gained per cubic metre of water is $\Delta\rho gh = 1.0 \times 9.8 \times 100 = 980$ J. If we obtained this energy from fossil fuel which produced 1 gCO_2 Wh^{-1} (3.6×10^3 J), the carbon dioxide produced would be 0.27 g. The 1 m^3 of water contains 294 mg of nitrogen and is able to dissolve an additional 0.84 gC before it is subducted. (The nitrogen is able to capture $294 \times 20.8 \times 10^{-3} = 6.1$ g of CO_2; but this amount of CO_2 just compensates for the carbon remineralised.) The energy needed is about 32% of the carbon captured. Lovelock and Rapley (2007) recently advocated using wave pumps to provide this energy and so avoid the fossil-fuel carbon penalty. Here the cost is in the form of capital and the quite significant maintenance required for anything deployed in the sea.

The component of the water artificially upwelled will in general be undersaturated with carbon dioxide for present atmospheric conditions, and the energy needed (at US$0.1 per kWh) will cost in excess of US$4.5 per tonne of carbon dioxide. It may be necessary to source the water from deeper in the thermocline and so the cost will rise rapidly both because of the extra lift and because of the greater density difference.

A Japanese research group, Toyota *et al.* (1991), raised up water in 1989/1990 from about 200 m depth. However, this did not produce a measurable increase in new primary production. Probably the deep water diluted too quickly for there to be a change in primary production measurable by the techniques used.

Iron fertilisation to utilise macronutrients

Fertilising with iron those regions of the ocean with an excess of macronutrients has attracted attention because iron is a micronutrient and only small quantities are required to stimulate photosynthesis and produce organic carbon. There are two regions in Figure 5.11 where the surface concentration of nitrogen is not low. These are the high latitude oceans and a band of the tropical Pacific. Nitrogen is available in these high-nutrient, low-chlorophyll (HNLC) regions because it is generally believed that a micronutrient is limiting phytoplankton growth. Light also limits phytoplankton growth in the polar regions.

Tests, starting with Martin *et al.* (1994), have been very successful in showing iron is a limiting nutrient in high-nitrogen regions. A 'massive phytoplankton bloom' was created by Coale *et al.* (1996) and reproduced by others. There is some progress in commercialising this technology of providing the micronutrient, e.g. Markels and Barber (2001). In the Southern Ocean, under the assumption that on returning to the surface the iron is again missing from the water, Caldeira (2003) found quite short retention times predicted by his model.

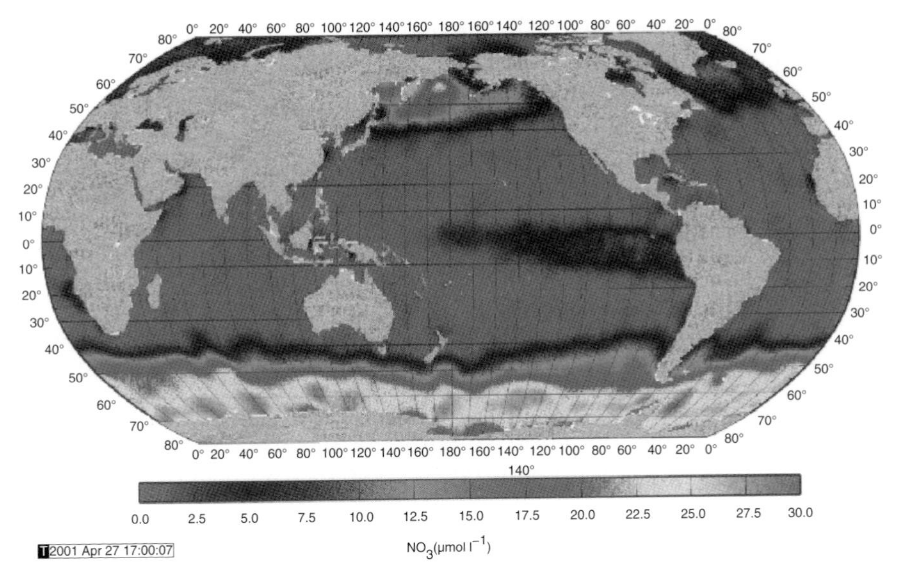

Figure 5.11 See plate section for colour version. Average surface nitrate levels in the upper ocean, after Levitus *et al.*, 1994.

In the Equatorial Pacific, models by Gnanadesikan *et al.* (2003) show that nitrogen used up by adding iron reduces new primary production elsewhere, a phenomenon known as nutrient stealing. At present, the excess nitrogen drifts westward across the Pacific and it presumably gains iron from the atmosphere. As can be seen in Figure 5.11, the Pacific equatorial concentration of nitrogen decreases to near zero by longitude 180°. Interest in iron fertilisation has led to more complex models, such as that constructed by Aumont and Bopp (2006). Since they assume the added iron is lost on remineralisation and that much remineralisation occurs at modest depth, the amount of extra carbon that can be stored in the organic cycle is limited by its rapid return to the surface ocean.

Economics has been considered by Ritschard (1992), where he suggested the cost of iron fertilisation was US$0.1 to 0.6 per tonne of organic carbon, say US$1 tonne of CO_2 (temporarily) stored. The stoichiometric molar ratio of carbon to iron is of order 10^6; however, the low efficiency of uptake makes a practical number much lower. Uptake efficiency may be less than 1% and this is believed to occur because iron attaches itself to particles that then sink out of the photic zone. The complexity of scavenging of iron means that, in an iron-limited situation, one cannot be confident that regenerative production follows new primary production. Without regenerative production, export production is only a small fraction of new primary production.

If the ratio of iron added to carbon stored is as low as 10^{-4} mole of carbon to a mole of iron, as suggested by Buesseler and Boyd (2003), the cost of the iron remains low. Let us use ferric chloride, molecular weight 162, as an example. A budget figure for ferric chloride is US$400 per tonne, which is 162/55 × US$400 per tFe = US$1178 tFe^{-1}. The carbon dioxide exported from the surface by a tonne of iron is $10^4 \times 44/55$ = 8000 tonne. We see again that even the more recent figures of the actual measured export of carbon (which is most likely low as Buesseler and Boyd (2003) point out since the growth of phytoplankton was still proceeding when they stopped the experiment) make the iron costs negligible, at about 1178/8000 = US$0.15 per tonne of CO_2. Delivery costs are also small. Say the iron injection point is 2000 km from port and at tanker steaming speed it takes 10 days port to port. Shipping costs might be of order US$0.2M per voyage. If 1000 tonnes of iron is injected, leading to an uptake of 8 $MtCO_2$, the transport cost per tonne of carbon dioxide taken from the upper ocean is US$200 000/8000 000 = US$0.03 per tonne. We must deduct the CO_2 generated by the production and delivery of the iron. The costly item is the spreading of the iron over the sea surface. Only a limited amount of carbon is exported because scavenging removes the iron quickly. This limits the reliance on diffusion to spread the iron, and so lines of injection must be close together. This in turn requires much steaming by the delivery vehicle.

By circulating more nitrogen via the biological pump, the production of nitrous oxide in the ocean might be enhanced. Away from the oxygen rich surface waters, there is a correlation between the nitrate concentration and the amount of N_2O. As the global warming potential of N_2O is some 300 times that of CO_2, the CO_2 equivalent, CO_{2e}, of the extra N_2O might be significant. Jin and Gruber (2003) found a large flux of N_2O because of the low efficiency in nitrogen usage as the retention time in HNLC polar regions is low. Increased primary production may also increase the release of dimethyl sulfide, DMS. The sulfur aerosols both scatter radiation and influence cloud formation. It has been suggested that DMS might reduce the net radiation on the earth and counteract the heat-trapping nature of N_2O. This suggests that the impacts of enhanced biological activity in the ocean will need to be assessed carefully.

To estimate the amount of carbon that might be stored by iron fertilisation, we will consider only the Southern Ocean. Peng and Broecker (1991) thought that 17×10^6 m^3 s^{-1} of nutrient-rich water flowed to the Southern Ocean surface layer during periods of adequate light for photosynthesis. If iron nourishment manages to convert an additional 2 μM (2×10^{-3} mol m^{-3}) of phosphate to organic matter and the Redfield ratio C : P is 106 : 1, then the organic carbon created is $17 \times 10^6 \times 2 \times 10^{-3} \times 106 \times 12$ g s^{-1} = 43.2 t s^{-1}. Over a year this is 1360 MtC yr^{-1} or 5 $GtCO_2$ yr^{-1}. This would be a useful contribution to climate management of about 20% of anthropogenic emissions at present. However, this organic carbon is only stored for a short while.

In summary, while iron fertilisation seems attractive, the uncertainty that it is permanent in the polar regions, or if it is merely shifting the region of ocean uptake in the equatorial regions, has delayed the introduction of commercial demonstrations. It may be that the enhancement of the marine protein production will provide justification for iron nourishment, even if its role in storing carbon dioxide is small.

Ocean Nourishment: cyanobacteria

In land-based agriculture, crops such as legumes are often planted in rotation to provide nitrogen to the soil. Cyanobacteria play the same role in the ocean by utilising di-nitrogen directly to form organic material. Moore *et al.* (2006) model how the bacterium *Trichodesmium* might respond to additional iron. They suggest that, over significant regions of the ocean, there is adequate phosphorus to allow the production of reactive nitrogen. With remineralisation of the cyanobacteria, the nitrogen fixed by the bacteria then enters the pool of preformed nutrients to support further primary production. Markels and Barber (2001) pointed this out under the heading of *ocean fertilisation*.

Doubt remains, as some culture-bottle experiments where the ocean water had adequate phosphorus found that additional iron was not enough to facilitate the growth of cyanobacteria. Organic nutrients seemed to be also needed.

Ocean Nourishment: macronutrients

Nitrogen, phosphate and silica, the macronutrients, can be injected into the ocean. There are many regions of the ocean that have limited phytoplankton growth due to the shortage of macronutrients. There is always adequate inorganic carbon. The nutrients in the upper ocean are refreshed by upwelling, and augmented by inputs from the rivers and the atmosphere. Phytoplankton consume these nutrients until one or more is exhausted. The availability of phosphorus, silica and trace nutrients can be assessed by considering the supply of these nutrients via upwelling and external input to the ocean. For example, there are estimates by Tiessen (1995) of the amount of new phosphate entering the ocean from the rivers.

The culture bottle experiments cited above support the contention that macronutrients are the limitation to further phytoplankton growth in much of the ocean. It is mostly nitrogen (Howarth, 1988) that is limiting. It must be recognised that, if enough macronutrient is added to the upper ocean, the phytoplankton will eventually exhaust the micronutrients. Then the addition of further nitrogen will be of no value until there is further upwelling. This however is not an important point because, just as Thomas (1970) did, one could add a broad spectrum of nutrients.

The point is that only a simple mix of macronutrients in some areas is sufficient to produce additional photosynthesis.

Phytoplankton are effective at capturing the inorganic carbon from the water and, by using solar energy, converting the carbon to organic matter and releasing the oxygen. The turbulent diffusion in the upper sunlit region of the ocean, where much photosynthesis occurs, ensures that there is a ready flux of carbon from the ocean surface. Here Henry's Law says that the partial pressures in the water and the atmosphere will try and equalise. Thus we have a process where some of the carbon dioxide (released from the flue stacks of power stations and other processes involving the combustion of carbon) is transported by the atmosphere to the sea. Here it crosses the surface to be converted to organic carbon by photosynthesis.

While new primary production remains constant, the organic carbon cycle returns as much carbon dioxide to the atmosphere (due to upwelling) as it exports from the upper ocean. By increasing the available macronutrients in the low and mid-latitudes, where there is ample sunlight, the ocean primary production can be enhanced. This in turn increases the organic carbon cycle and sequesters carbon in the oceanic part of the cycle.

This observation allows us to make a first-order estimate of cost of sequestering carbon in the sea by Ocean Nourishment using nitrogen. Nitrogen is the nutrient of greatest mass in phytoplankton. Reactive nitrogen is produced today from the atmosphere by a widely practiced technology invented by Haber and Bosch. The prices of compounds that contain reactive nitrogen vary with the cost of inputs and with a periodic nature due to the long investment cycle needed to overcome under supply. Consider urea, NH_2CONH_2, as a convenient form of nitrogen. Let us assume urea is available at the coastline at US$200 per tonne of nitrogen. As we saw above, about seven tonnes of phytoplankton contains about six tonnes of carbon and one tonne of nitrogen. For the present simple analysis, you can see that converting one tonne of carbon to organic carbon needs about US$200 / 6 = US$33 of chemical. Recalling that the ratio of C to CO_2 is 12 to 44 by mass, the chemical cost is of order US$33 / 3.7 = US$10 per tonne of CO_2. This is a very attractive order-of-magnitude cost.

Increasing the primary production by any of the above is expected to increase the marine biomass. One hypothesis is that recruitment success determines the species distribution of fish such as small pelagics, while food supply determines the total biomass. Ware and Thompson (2005) showed a strong correlation between chlorophyll concentration and long-term fish catch.

Macronutrient delivery to the ocean and its cost

Figure 5.12 shows a conceptual land-based factory fixing nitrogen from the atmosphere and enhancing photosynthesis at the edge of the continental shelf. At the core

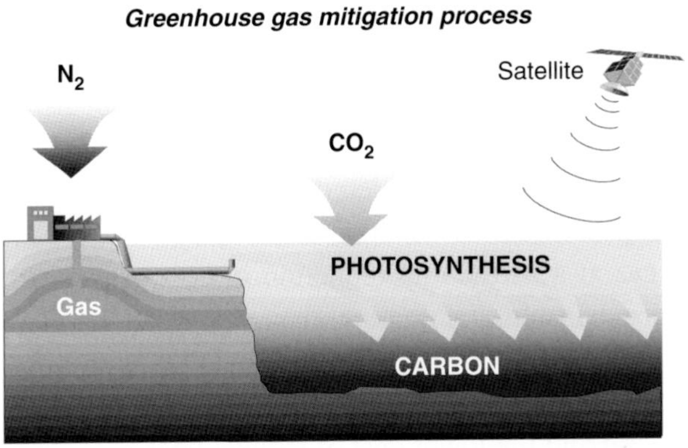

Figure 5.12 Schematic of the Ocean Nourishment concept examined in Jones and Otaegui (1997).

of the system is the production of the macronutrient, reactive nitrogen, in the form of ammonia, NH_3, synthesised from methane, CH_4. The process of manufacturing ammonia produces carbon dioxide, and this should be subtracted from the amount of atmospheric carbon sequestered by the introduction of nitrogen to the upper ocean. Fortunately it is possible to combine this carbon dioxide with the ammonia to form urea. So little carbon dioxide needs to be emitted to the atmosphere. The reactive nitrogen (and possibly other nutrients) is introduced to the surface water of the ocean where the standing stock of phytoplankton will use the nutrients to produce organic matter.

The process should be designed to provide the nutrient at a depth and with an initial dilution that ensures uptake by the standing stock of phytoplankton before their concentration becomes undesirably high. Injection of the nutrient at too great a depth will place it out of reach of the phytoplankton. The initial uptake of nitrogen might not be complete, and so it is prudent to assume an uptake efficiency of 70% of that supplied. Some of this allowance can be for a host of other factors that lower storage efficiency.

Measuring the actual carbon uptake can be achieved through regular retrieval of appropriate satellite images for the area. These images will measure the amount of chlorophyll at the surface of the ocean which, combined with estimates of the ocean mixed-layer depth, provide a measure of the new primary production induced downstream of the injection point. In addition, regular surveying carried out by vessels will show the reliability of this form of estimation of carbon dioxide uptake for the Ocean Nourishment scheme.

Using the satellite information, the photosynthesis process can be controlled by adjusting the flow rate of nutrients and/or changing the initial rate of mixing. One imagines being able to increase or decrease the number of exit ports on the vertical risers. It is the control of the process that reduces the risk of producing harmful or undesired changes in the ocean downstream of the nutrient release point.

The example used above involved nourishing the ocean using a shore-based nutrient manufacturing plant and a subsea pipeline for delivery of the nutrient to the photic zone. An alternative considered by Jones and Cappelen-Smith (1999) was to use a floating nourishment plant. One advantage would be the ability to use stranded natural gas at a substantial reduction in cost of production of nutrient. When the gas was exhausted, the floating plant would be moved. A floating nutrient plant can be expected to have a higher capital cost than a fixed, shore-based plant but it does not require a subsea pipeline. The main advantages are flexibility of the use of the asset and the lower cost of stranded gas.

Another alternative is to deliver the nutrient by ship rather than through a fixed pipeline. Here the nutrient, possibly urea, is loaded on a ship at a wharf of an ammonia manufacturing facility and taken to the release site, where it would be broadcast or pumped into the photic zone (Judd *et al.*, 2008). The ship would then return to the wharf to load more nutrient.

Ocean thermal energy conversion (OTEC) could be used to make a completely self-sustaining system. OTEC derives energy from the temperature gradient in the sea and could be considered a renewable energy source. This energy can be used to produce hydrogen from seawater and combine it with nitrogen from the air. As well, there is the nutrient-rich wastewater. This suffers the difficulty of being carbon rich, as discussed above. OTEC is a non-established technology and is unlikely to be used in Ocean Nourishment in the short term.

Reactive nitrogen is available in almost unlimited amounts from the atmosphere using the Haber–Bosch process. The impact on food agricultural production has been dramatic, and Haber was awarded a Nobel prize for his part in the discovery of the process. The cost of manufacture, however, remains high. One of the attractions of using cyanobacteria to produce reactive nitrogen is that the cost promises to be low.

Some properties of a modern nitrogen-fixing plant using natural gas are listed below:

Ammonia production	2000 t day^{-1}
Natural gas consumption	30 GJ $tonne^{-1}$ NH_3
Freshwater consumption	1000 kg $tonne^{-1}$ NH_3
Electricity consumption	20 kWh $tonne^{-1}$ NH_3
Labour for operation	6 persons per shift × 4 shifts
Labour for maintenance	6 persons per shift × 4 shifts
Operating days per year	330
Yearly NH_3 production	660 000 tonnes

The costs in US dollars we have assumed for the construction of a 2000 t day^{-1} NH_3 plant at a greenfield location are:

Capital cost	\$300M
Natural gas costs	Variable

Other costs in US dollars per tonne NH_3:

Direct labour	\$2
Maintenance	\$2
Catalyst	\$3
Electricity	\$2
Cooling water	\$2
Testing	\$0.2
Freshwater	\$0.1
Total	\$11.3

In addition, the nourishment delivery system of a subsea pipeline has been estimated to have a capital cost of US\$50M, and it can be operated by the same staff that are involved in the production of NH_3. Thus we have a total capital cost of US\$350M.

On a yearly basis, the expenses are for 19 800 000 GJ of natural gas, interest on US\$300M debt, return on equity and (other) operating costs of US\$7.46M. Using the figure from Jones and Otaegui (1997) of 12 t of CO_2 sequestered per tonne of nitrogen provided, a yearly supply of 7.92 Mt of CO_2 carbon credits would be provided to the market. The ratio of 12 : 1 carbon dioxide to ammonia assumes a Redfield ratio for phytoplankton and allows for the carbon dioxide produced in the ammonia synthesis process. The value of the extra fish produced is assumed to be not captured by the operators of the nourishment plant. This is a conservative approach, as it seems very possible that extra fishing rights could be auctioned, as a number of countries have such fish quotas in place. Shoji and Jones (2001) provide a more detailed analysis.

The capital requirement can be reduced by using ship delivery of the nutrient rather than constructing a greenfield nitrogen plant. Now the equity required can be quite modest, by leasing the ships and buying the urea from the most economical source. However this process is likely to lead to higher costs per tonne of carbon dioxide stored.

Jones and Altarawneh (2005) looked at the economics in the future of using coal as a feedstock rather than natural gas. While coal is a more expensive option today, its ready availability will make it potentially more economical in the future than natural gas, the price of which is rising steadily. Coal, as an alternative source of abundant energy, places a cap on the future cost of Ocean Nourishment.

It is assumed that additional costs of any other nutrients required will be small. They would be used to supplement the reactive nitrogen and only used if cost effective. Phosphorus is available in some ocean regions, but in others it would need to be supplied by mining. According to Marinov (2008), 4230 GtC could be stored in the ocean if one waited for this available phosphate to be upwelled to the photic zone, a process which will take many hundreds of years. For more rapid storage of carbon, phosphate would need to be obtained by land-based mining.

For iron fertilisation, ship delivery would appear to be the best option because of the small mass of iron needed and the long distance from the source of iron.

The carbon credits discussed here are in terms of the equivalent impact of not releasing one tonne of carbon dioxide into the atmosphere. It is not a reduction in atmospheric carbon dioxide of one tonne. This complexity comes about because the introduction of an additional tonne of carbon dioxide into the atmosphere does not lead to all of it residing in the atmosphere. Some of the carbon released by fossil fuel burning will find its way into the ocean by Henry's Law. Some will enter as a result of Ocean Nourishment-induced photosynthesis. Some will enter as a result of previously introduced nutrients being recycled from below the photic zone, but these are expected to have a compensating amount of carbon released back into the atmosphere. Modelling by Orr and Sarmiento (1992) has looked at the long-term effects of Ocean Nourishment on atmospheric concentrations of carbon dioxide, but has not expressed the results in a form suitable to assess the uptake due to adding nutrient. A more recent report by Matear and Elliott (2004) addresses a more relevant problem, where recycling of the purposefully added nutrients is considered. A number of effects reduce the long-term storage efficiency. These include the formation of calcium carbonate shells, and the fact that some of the carbon, while trapped in the organic carbon cycle, resides in the atmosphere. Jones and Caldeira (2003) directly addressed the question of long-term sequestration under the restrictive assumption that nitrogen loss was negligible, and found retention times well in excess of 500 years.

Macronutrient nourishment appears to be economically attractive when compared with other large-scale carbon dioxide mitigation strategies, and has the advantage of contribution to the global food supply by enhancing primary production of the world's oceans.

Remote sensing of carbon uptake

Remote sensing of the ocean from space has increased our understanding of the ocean (Jones *et al.*, 1993), and satellite remote sensing will be able to contribute to monitoring ocean fertilisation activities. Already, ocean colour sensors have been

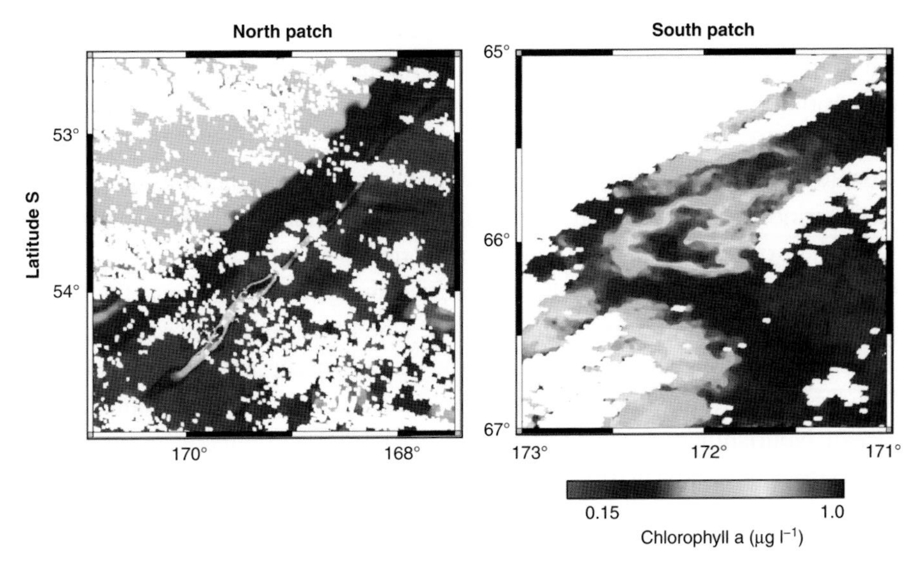

Figure 5.13 See plate section for colour version. Ocean colour images showing increased chlorophyll in the enriched patches as red, after Coale *et al.* (2004).

used to track the increase in chlorophyll as a result of an injection of iron in the Southern Ocean, as shown in Figure 5.13. It will be desirable to monitor the uptake of carbon dioxide by the new primary production that results from the introduced nutrients. From such information primary production can be inferred, for example as in Balch *et al.* (1992) or Pan *et al.* (2005).

Ocean colour satellites rely on the sunlight backscattered from the upper ocean to detect the concentration of scatterers. This can be the chlorophyll of phytoplankton or the scattering of suspended sediments. In so-called Case 1 waters, the water-leaving light is dominated by chlorophyll, and estimation of the concentration of chlorophyll is more reliable. As remote sensing would appear to be useful for Ocean Nourishment carried out in relatively barren waters, Case 1 analysis would seem to be sufficient. Remote sensing is a powerful tool to ensure that Ocean Nourishment is not causing eutrophication.

Benefits and risks

Storing carbon in the ocean rather than leaving it in the atmosphere has a number of benefits and creates a number of risks. A benefit of all the options for intervention discussed in this book is that climate change damage is reduced. In the surface ocean, Caldeira and Wickett (2003) have shown that the burning of fossil fuel, and leaving the carbon dioxide in the atmosphere, will decrease the pH, disadvantaging calcifying organisms.

Capture and direct injection places the pH change in the deep ocean, away from much of the biological activity of interest to humans. Dilution will ensure a lower concentration of acidity than direct invasion into the much smaller volume of the surface ocean. Taking the next step of neutralising the carbon dioxide with calcium carbonate overcomes the above difficulty of direct injection, and provides benefits to the marine environment. However there is a substantial economic penalty in moving the extra material (2.3 t $CaCO_3$ per tonne of CO_2).

Ocean Nourishment changes the organic carbon cycle but mitigates the increase in surface acidity. Similar to direct injection, when the remineralised organic matter exported from the surface ocean returns to the surface, the surface pH will be lowered. However, this is many years later and the acidity is much diluted.

Ocean Nourishment is able to capture carbon that is already emitted to the atmosphere. For some distributed sources of carbon dioxide such as aeroplanes (estimated emission in 2030 of one gigatonne of CO_2 per year) it will be very difficult to avoid emitting carbon dioxide. Another benefit of Ocean Nourishment is that it can convert the waste product, carbon dioxide, to marine protein. The fish produced can be considered a free good.

There are risks associated with ocean storage of carbon. Over-aggressive application of Ocean Nourishment might draw down oxygen levels in the deeper ocean, increase the production of N_2O and so lower the efficiency of the greenhouse gas sink below that assumed. Increased export of organic matter from the surface ocean due to enrichment of the surface waters will consume oxygen in the ocean thermocline as bacteria turn organic carbon back to inorganic carbonates. The water in the thermocline gained its oxygen when it was at the ocean surface. If the length of time this water is exposed to the enhanced rain of debris is long, compared with the time for the thermocline water to recycle to the surface and be refreshed with oxygen, then a reduction in oxygen will be experienced. This may well place stress on the ecology of this region of the ocean, and if drawdown of oxygen was large, it might also increase the formation of N_2O, since it is believed that anoxic conditions lead to increased denitrification. Harmful algal blooms (HABs) are undesirable and need to be cut off from their supply of nutrients if they have the prospect of blooming.

Carbon capture and ocean storage will change the properties of the cooling water plume and may over-stimulate the growth of shellfish in the plume path.

It is feasible to carry out ocean storage with known technology. Direct injection can exploit deep-sea drilling technology. Using cooling water in CCOS with neutralisation is a low-technology activity. Mining and grinding calcium carbonate has been practised for years. Subsea transport of liquids and the production of reactive nitrogen are mature technologies. Thus the technical risks in providing the nutrients to the surface ocean are low. Uncertainty exists in the biological response, and this justifies a pilot experiment. Shipping carbon dioxide is closely related to the transport of LNG. Here the risks are in the deep water injection pipe.

There are, in addition, uncertainties to do with the public acceptability of manipulating the ocean. In the affluent countries where food security is not a concern, there are groups who are opposed to using the oceans to 'solve a waste problem' even though Ocean Nourishment would increase the supply of protein needed to feed the world's rising population. A model of how the enhanced food chain might operate is presented in Jones (2004). It is possible that, while the *Green Revolution* fed the recent two-billion increase in population, the next two-billion increase might have to rely on a *Blue Revolution* for their protein.

Carbon sinks can allow us to retain the existing fuel infrastructure while producing no net greenhouse gas. Coal-fired electricity is one of the most flexible and low-cost forms of energy. By burning coal at maximum efficiency in large generating plants with adequate emissions control and taking up the carbon dioxide in the ocean, a pollution-free form of energy can be provided. Low-cost energy has been one of the keys to providing an affluent society. There is capacity in the ocean to sequester carbon from the next century of fossil fuel use. There are adequate supplies of fossil fuel and, with a century horizon, the poverty in the developing world may be overcome with the aid of plentiful energy. (Of course it is not just access to energy that handicaps the poor.)

Comparison of ocean sequestration techniques

Tokimatsu *et al.* (2004) used the climate damage from a Nordhaus model to calculate the benefit of direct injection of carbon dioxide. They argued that actions that reduce climate damage produce a benefit greater than the cost of direct injection into the ocean. Jones and Altarawneh (2005) followed a similar approach for Ocean Nourishment. Both studies show benefits exceeding the costs assumed in the Nordhaus model. As well as reducing climate damage, both Ocean Nourishment and alkalinity shifting have secondary benefits for marine life. These benefits were not taken into account in the above economic discussion of climate damage.

The costs of the three processes of ocean storage can easily be ranked. Capture and direct injection into the ocean is the most expensive because capture is an energy-intensive activity. This is discussed further in Riemer (1996) and in Chapter 6 as part of the discussion on capture and geological sequestration. Changing the Total Alkalinity involves handling large masses of solid material (limestone), but capture of small amounts of carbon from concentrated sources (e.g. flue gases) can be achieved with little energy penalty. This opportunity is limited by the amount of circulating seawater (used for cooling). New coastal power stations do offer an attractive option for CCOS. Ocean Nourishment has the lowest cost, as most of the energy is provided directly by the sun (photosynthesis), and the introduced nutrients are only a fraction of the mass of the carbon sequestered. If cyano-bacteria can be harvested reliably to produce inorganic nitrogen, this promises to be the cheapest form of ocean fertilisation. However, macronutrient nourishment would appear to be more reliable, with our present understanding of ocean processes.

Maintaining the thermohaline circulation

We finish the chapter with a thought, not about changing the solar radiation trapped by the earth–ocean–atmosphere system, but on stopping the redistribution of heat by the ocean.

The thermohaline circulation may be slowing due to less dense waters forming near the poles, due to both temperature rise and more ice melt. Water carried away from the surface is nearly saturated with carbon dioxide and is close to being in equilibrium with the atmosphere. Slowing of the overturning circulation will have the opposite effect to the induced upwelling discussed above. Less old water with low carbon content will reach the surface ocean. If it is not practical to increase the concentration of carbon dioxide being carried to the deep ocean, we are left with trying to maintain or increase the flow.

Let us assume that we find the polar waters are 2 °C warmer as a result of climate change. Thus 2×4.186 kJ m^{-3} of heat energy needs to be removed to restore the status quo. Taking the thermohaline circulation as 280 Sv, we need to remove 280×8.37 GJ s^{-1} = 2300 GW of energy. As a large modern power station produces 1 GW of power, this is of the same order as the total global installed power production (Appendix 5). Making ice and transporting the ice to the region of sinking water is maybe the best method we can think of.

Damming the mouth of the Mediterranean is another idea to mitigate regional climate change. The exchange of water between the Atlantic Ocean and the Mediterranean is believed to influence the climate of Northern Europe, and so controlling the outflow might allow climate intervention.

Exercises

Exercise 5.1 How much carbon dioxide would be generated by transporting carbon dioxide by ship from the shore to a disposal site some 100 km away in deep water?

Exercise 5.2 If one wished to sequester 1000 t yr^{-1} of carbon dioxide from a coastal power station burning black coal, what sized Ocean Nourishment plant would be required?

Exercise 5.3 The capture of carbon dioxide from flue gases is expensive. Could this cost be avoided by compressing the whole of the flue gases and pumping them into the deep ocean?

6

Increasing land sinks

The previous chapter discussed the 70% of the globe covered by the ocean, and in this chapter we consider the remaining 30% that is land. While the terrestrial ecosystem provides less than a tenth of the carbon storage of the ocean, it is about as active on a seasonal basis in terms of carbon flux in and out of the atmosphere. The upper 1 m of soil contains some 2000 GtC, while the present land vegetation stores about 750 GtC, of which about 300 GtC is stored above ground in forests. These are large stores of carbon which can be both enhanced as an alternative sink for atmospheric carbon or mobilised (unintentionally) to produce additional emissions. Most of the fossil fuel that is being burned to produce the rising atmospheric carbon dioxide came from the land. Remember that we estimated the recoverable fossil-fuel reservoir as 7000 GtC. Logging of the forests for land clearance, and other changes in land use, add to the carbon dioxide emissions to the atmosphere at a rate of 2 GtC per year. Vegetation grown for food is cycled once or twice a year and so does not hold much of the mobile carbon. In this chapter we will look at the three approaches to land storage of carbon.

The most direct approach is to pump compressed carbon dioxide into depleted oil and gas wells and this is already done for the purpose of recovering more oil. Aquifers can also be used. The carbon now being released into the atmosphere was sequestered in the earth's crust before the human race began to liberate it to provide the energy to drive an industrialised society. It seems attractive to return the waste product, carbon dioxide, to the ground. A second concept is to convert carbon dioxide to a solid material such as magnesium carbonate which can be used as a form of landfill. Finally there is the prospect of changing the biological storage, either through farming practice or through forest generation or management.

In this chapter we will only briefly treat forest and agricultural practices, as they represent a specialisation that is treated elsewhere more fully than we can achieve here. As well, the details are not usually an area of expertise of engineers. Readers are referred to Metz *et al.* (2001).

Terrestrial primary production

Carbon from the atmosphere is converted to organic carbon by plants using photosynthesis. The net photosynthesis, the difference between uptake and respiration, is stored in the roots, trunks and leaves of plants. Net primary production of vegetation (photosynthesis less respiration) is not the ecosystem flux of carbon to the atmosphere, since respiration is not taken to mean the release of CO_2 from the litter or the burning of biomass. Net primary production of vegetation depends upon variables such as rainfall, temperature and sunlight, and models for the primary production can be developed using vegetation indices generated by satellite observations. Nemani *et al.* (2003) found an increase in net primary production over the last 20 years of about 6%, and found that this increase occurred mainly in the tropics and secondarily in high northern latitudes (Figure 6.1). During the same 20 years, the temperature rose about 0.3 °C and the carbon dioxide concentration by 9%. It was the CO_2 fertilisation and changes in the tropical cloud cover which accounted for the net primary production changes.

It is believed that increased concentration of carbon dioxide allows greater primary production because, we assume, the transfer of carbon into the leaves is a limiting process in many plant varieties. In the ocean, in contrast, carbon dioxide is abundant and it is nutrients that control primary production in most situations. However, Nemani *et al.* (2003) conclude that CO_2 fertilisation is not the whole cause of the increase of net terrestrial primary production. This increased net primary production has averaged 0.3 GtC per year, which over 20 years is comparable with the 6 GtC yr^{-1} of anthropogenic emissions.

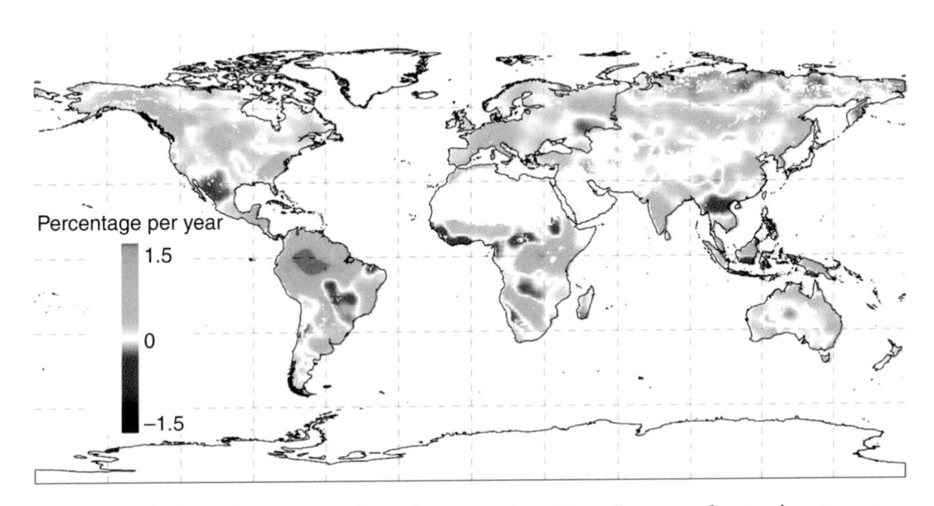

Figure 6.1 See plate section for colour version. The changes of net primary production calculated from a productivity model, after Nemani *et al.* (2003).

A larger net primary production does not necessarily imply sequestration of carbon dioxide from the atmosphere. Mature stands of vegetation are in steady state on timescales of years, as decomposition of litter and the loss of carbon from the soil returns carbon to the atmosphere as fast as photosynthesis extracts carbon. In the case of land clearance, the terrestrial vegetation can release previously sequestered carbon into the atmosphere.

Humans are responsible for about half the net primary production of the land surface. Agriculture occupies some 20% of the land, and forestry occupies some more. Land net primary production is about 60 GtC yr^{-1}, while the ocean net primary production is about 50 GtC yr^{-1}. As net primary production is the base of the food chain, can we do more to accelerate this energy conversion process and produce more food? Quite possibly, but this should not be confused with the storage of carbon. Note that both these numbers are averaged over the seasonal recycling of carbon.

Growing trees

The 20% of land used for agriculture once was covered with trees, and their removal released carbon. The total land area of the world is $148 \times 10^{12}\,m^2$, and so 20% is $29.6 \times 10^{12}\,m^2$. If we assume that the original density of aboveground carbon was $0.025\,tC\,m^{-2}$, then the amount of carbon that has been released, preparing the land for agriculture, is $740 \times 10^9\,tC$. This is larger than that released by fossil fuel burning and of the same order as the present store of carbon in the atmosphere. Of course the present carbon in the atmosphere is not just from land clearing. Much of the carbon from historic land clearing is now in the ocean and in the remaining forests.

Winjum *et al.* (1992) estimated forested land at 4.08×10^9 hectares ($40.8 \times 10^{12}\,m^2$). A hectare is 10 000 m^2 or 2.47 acres. These forests which cover some third of the land surface of the earth hold about 90% of all aboveground terrestrial carbon. Forests hold about 700 GtC above and below ground and are being cleared at the rate of 13×10^9 ha per year. The IPCC estimated the land plausibly suitable for conversion to forestry might be 560 Mha ($5.6 \times 10^{12}\,m^2$), but this statement involves many qualifications. The amount of carbon that might be stored in forests is of order 200 t ha^{-1}, and so the storage potential is 112 GtC. The average number of trees and their size are the main determinants of the amount of carbon that a forest contains. Thus forestry might be able to provide a sink for the next 15 years of IS92a emissions, or for 7 years emissions around the year 2100.

In steady state, the remaining forests do not sequester carbon. However if agricultural or pastoral lands are planted as forest, the new trees store carbon in their trunks and foliage. As well, their root system contains carbon as does the surrounding soil. Trees must be considered a temporary store of carbon because of the potential pressures on future generations to convert forestlands to other uses and release the carbon back to the atmosphere. As the CO_2 concentration in the atmosphere increases, we expect the carbon stored in trees to rise, since the uptake of carbon dioxide is one of the constraints on plant size. However, the studies reported by Davidson and Hirsch (2001) suggest that the nutrient supplies quickly provide another limitation.

In an analogy to Ocean Nourishment considered in the previous chapter, providing fertiliser to existing forest can in most cases also increase the carbon stored. The carbon to nitrogen ratio in trees is higher than for phytoplankton. However, in contrast with Ocean Nourishment, this is not an on going process. Once a region is provided with adequate fertiliser and the forest becomes mature, a new steady state is reached. Extra fertiliser then leads to 'nitrogen saturation'.

This is not to rule out cutting of timber by rotation, but there is negligible storage achieved. What storage there is comes from long-life timber products. Incorporating the mature wood into buildings and the like delays this process of decay and the return of the carbon to the atmosphere, but does not eliminate it. Newell and Stavins (2000) suggest that harvesting, despite the income from the sale of lumber, increases the cost of carbon sequestration in forests. This occurs in their model, which examines the present net value, because the carbon is sequestered more quickly in the no-harvest scenario, as can be seen in Figure 6.2. (Look at the present net value discussion below and in Appendix 2.) There is no further uptake of carbon after 100 years or so.

An attractive alternative is to use the forest products as fuel, as a substitute for fossil fuels. Forest products have a much lower energy density than coal; so much larger volumes must be handled.

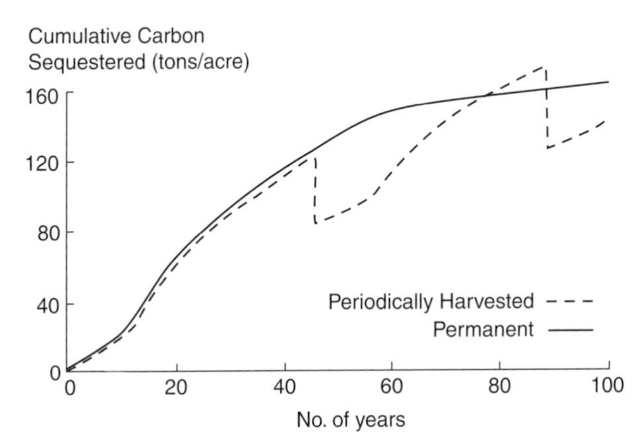

Figure 6.2 The carbon sequestered as a function of time for Loblolly Pine, after Richards *et al.* (1993). Note acres not hectares.

New forests imply taking land out of agricultural use. With the relentless growth of the world population, the pressure to increase the amount of agriculture is high. We have a conflict between increasing the food supply (by using present practice but more land) and increasing the store of carbon in forests. In fact the area of forest is declining due to land clearance. This can be expressed as the (lost) opportunity cost of forest sequestration.

Another issue to be considered is the change in albedo induced by increasing the forestland. At high latitudes, additional trees, being dark green, can reflect less solar radiation than white snow. In the tropical regions this is not important, and so reforestation in the tropics is more promising.

Estimating the cost of carbon sequestration by forests is not a simple matter. The sequestration of carbon occurs over time, as Figure 6.2 illustrates. For this example of Loblolly Pine, the uptake over the first 40 years is about 247 t ha^{-1} (100 tC $acre^{-1}$), which is a rate of 6 tC ha^{-1} yr^{-1}. The simplest approach is to treat all the carbon stored over the growth period as sequestered at time 0. This is a poor measure. Another approach is to estimate the discounted present *amount* of carbon stored. A discount rate must be agreed upon. Often real (inflation-adjusted) values of 5% pa are considered reasonable discount values. The cost of carbon sequestered over time can then be expressed also as a present value.

Notice in Table 6.1 how the *discounted present amount* depends on the discount rate, and 'no harvest' gives more 'present value' carbon sequestration than periodic harvesting. While it is difficult to agree on a discount rate, it is a sensitive variable in estimating the present amount of carbon.

Like the discounted present amount of carbon stored, the *present value costs* can be estimated. The costs of forestation depend upon the amount of land converted to forests. A small amount of additional forest can use marginal agricultural land at low cost. In the study by Newell and Stavins (2000), they considered a region of the USA of approximately 13 million acres (5.3 Mha). This region is 0.6% of the land area of the USA (916 Mha). They enquired of the costs to induce landholders to increase the forested area. Carbon dioxide emissions from fossil fuels (in the year 2000) in the USA are 1500 MtC yr^{-1}.

How can this be expanded to get some feeling for the role forestation might play in mitigating the greenhouse gas concentration? For example, consider the case when 0.65 Mha of the 5.3 megahectares studied are changed to forest in year one. The annualised amount of carbon is calculated by discounting future storage (Table 6.1) so that the diminishing rate of storage with time (Figure 6.2) is not very important. Table 6.2 shows that this action stores carbon at an average cost of US$13.38 per tonne. The carbon stored is 2.4 MtC yr^{-1} for each of the next 90 years, and this is about 0.16% of the emissions of CO_2 from the USA (in the year 2000). To induce farmers to make this change from agriculture to growing forests,

Table 6.1. *Present value of the amount of carbon stored (tC ha^{-1}) in a new forest.*

Discount rate pa (%)	2.5	5	7.5	10
Natural regrowth of mixed stand				
Periodic harvest	153	107	76	23
Pine plantation				
No periodic harvest	200	123	85	62

Reproduced from Newell and Stavins (2000).

a subsidy of US\$13.38 tC^{-1} (US\$3.65 per tonne of CO_2) is required. Larger sinks of carbon dioxide cost more, so the average price of achieving a 0.28% reduction of current USA emissions is US\$20.3 tC^{-1} (US\$5.56 per tonne of CO_2). To sequester 10% of USA emissions (150 MtC yr^{-1}) would require about 38 times as much land (25 Mha). Twenty-five megahectares of land in the USA converted to new forests in the face of the continuing need for food production might drive up the opportunity cost of converting from agriculture. However, the price is attractive. The present net value assumes that, if the forest is cut down in the distant future, this is of little consequence since the discount rate (see Appendix 2) ensures almost no loss in present value. In reality the carbon is returned to the atmosphere and negates the storage.

Estimating carbon in forests

The carbon stored in a forest can be divided into four pools: carbon in aboveground biomass of live trees; carbon in woody and other debris; carbon in root systems; and soil carbon. We will discuss only the first. There are many methods that can be used to estimate the aboveground carbon in trees. The simplest is to use established correlations between tree trunk diameter and total biomass in the tree. Snowdon *et al.* (2000), for example, provide diagrams such as Figure 6.3, relating biomass per hectare to the area of tree trunks per hectare of land. This is known as basal area and is usually determined by measuring, at breast height, the circumference of the tree trunks in a sample area.

The fraction of the tree's biomass that is carbon, above and below the ground, next needs to be estimated in order to calculate the amount of carbon temporarily sequestered.

Costs of permanent forest sequestration

The cost of producing a forest can be estimated with reasonable certainty. However the costs involved in providing long-term storage of carbon in trees are much more problematic. An upfront cost is incurred in purchasing agricultural land for

Table 6.2. *Storage costs for harvested forests at 5% discount rate, after Newell and Stavins (2000).*

Increase in forested area (acres)	Increase (Mha)	*Present* amount of carbon (MtC)	*Annualised* carbon sequestered (MtC yr^{-1})	Percentage of one year of USA emissions[a] (%)	Cost to convert to forest (US$ $acre^{-1}$ yr^{-1})	Marginal cost (US$ tC^{-1})	Percentage of land newly forested (%)	Percentage of USA land area
518 000	0.21	21.2	0.784	0.05	10	6.61	4	0.02
1057 000	0.43	43.4	1.600	0.17	20	9.97	8	0.04
1615 000	0.65	65.7	2.445	0.16	30	13.38	12	0.07
2192 000	0.89	89.9	3.3	0.22	40	16.81	17	0.01
2787 000	1.13	114.1	4.2	0.28	50	20.27	21	0.13

[a] Annualised emissions / total USA emissions in 2000.

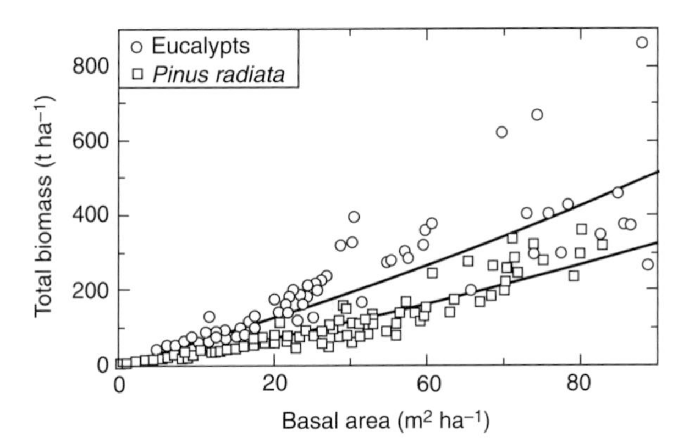

Figure 6.3 Relationship between total aboveground biomass and basal area, after Snowdon *et al.* (2000).

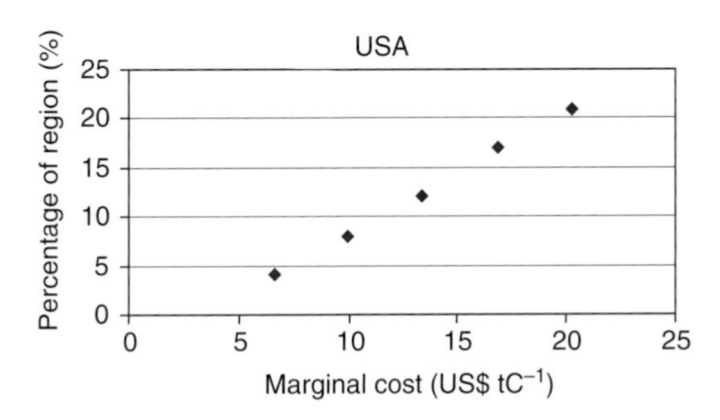

Figure 6.4 Marginal cost of converting farmland to forestry in the USA, see Table 6.2.

forestation. The sequestration comes over the next 40 years or so. Using present value comparisons, the fact that carbon storage stops is of little import. However, because of the long lifetime of carbon dioxide, one wants to know the cost of storage forever, as well as the present value. This could be expressed as the present value of an insurance policy or bond. Even a bond for 100 years might be quite expensive, increasing the prices proposed above. If the forest is then cleared or suffers loss due to a fire, what will be the status of the carbon credit? Ignoring this question, the costs discussed above can be seen in Figure 6.4. They can be thought of as the present-day costs of convincing landowners to convert their land to forests and harvest the trees, rather than earning income from farming.

Role of forests in mitigation

The following summary has been adapted from the IPCC Report *Climate Change 1995* (Houghton *et al.*, 1996). We have retained their classification of the confidence one can place in each assessment.

Three categories of promising forestry practices that promote sustainable management of forests and at the same time conserve and sequester carbon are:

(1) management for conservation of existing carbon pools in forests by slowing deforestation, changing harvesting regimes and protecting forests from other anthropogenic disturbances;
(2) management for expanding carbon storage by increasing the area and/or density in native forests, plantations, under agroforestry or storage in wood products; and
(3) management for substitution by increasing the transfer of forest biomass carbon into products such as biofuels and long-lived wood products.

We need estimates of the quantities of carbon that can be conserved or sequestered and the associated implementation costs of forest sector mitigation strategies. We also need to identify the limits on the amount of lands available for such mitigation strategies.

- The most effective long-term (>50 years) way in which to use forests to mitigate the increase in atmospheric CO_2 is to substitute fuelwood for fossil fuels and for energy-expensive materials. However, over the next 50 years or so, substantial opportunities exist to conserve and increase the carbon storage in living trees and wood products (*high confidence*).
- Under baseline conditions (today's climate and no change in the estimated available lands over the period of interest), the cumulative amount of carbon that could potentially be conserved and sequestered over the period 1995–2050 by slowing deforestation (138 Mha) and promoting natural forest regeneration (217 Mha) in the tropics, combined with the implementation of a global forestation programme (345 Mha of plantations and agroforestry), would be about 60 to 87 GtC (1.1 to 1.6 GtC yr^{-1}) – equivalent to 12–15% of the projected IPCC IS92a scenario (see Figure 1.8) – cumulative fossil fuel carbon emissions over the same period (*medium confidence*).
- The annual carbon gain from the above programme would reach about 2.2 Gt yr^{-1} by 2050. The gradual increase over time occurs because of the time it takes for programmes to be implemented and the relatively slow rate at which carbon accumulates in forest systems (*medium confidence*).
- Uncertainty associated with the carbon conservation and sequestration estimates is caused mainly by high uncertainty in estimating land availability for forestation and regeneration programmes and the rate at which tropical deforestation can actually be reduced; estimates of the net amount of carbon per unit area conserved or sequestered under a particular management scheme are more certain (*high confidence*).

- The tropics have the potential to conserve and sequester the largest quantity of carbon – 45 to 72 Gt – more than half of which would be due to promoting natural forest regeneration and slowing deforestation. Tropical America has the largest potential for carbon conservation and sequestration (46% of the tropical total), followed by tropical Asia (34%) and tropical Africa (20%). The temperate and boreal zones could sequester about 13 Gt and 2.4 Gt, respectively – mainly in the United States, temperate Asia, the former Soviet Union, China and New Zealand (*medium confidence*).
- The cumulative cost is uncertain. Land costs and other transaction costs will be substantial. The direct costs of only the forest practices to conserve and sequester the above amounts of carbon range from US\$247 billion to US\$302 billion at a unit cost of about US\$2 to US\$8 tC^{-1}. Transaction costs may significantly increase these estimated costs. These costs could be several times higher if land and opportunity costs and/or the costs of establishing infrastructure, protective fencing, training programmes and tree nurseries were included. No complete cost estimates are available (*low confidence*).
- Costs per unit of carbon sequestered or conserved generally increase from low to high latitude nations (*high confidence*) and from slowing deforestation and promoting regeneration to establishing plantations (*low confidence*). The latter trend may not hold if transaction costs of slowing deforestation are excessive.

Agricultural practices

There is pressure to clear more land for agriculture to supply the increasing demand for food. In the next 30 years there are expected to be two billion more people to feed. At present about 13 $Mha\,yr^{-1}$ is being cleared, and this releases about 1.1 $GtC\,yr^{-1}$ to the atmosphere. A halt to land clearing would mean a reduction of some 15% in carbon dioxide emissions. Thus land clearing is currently a significant source of anthropogenic greenhouse gas emission to the atmosphere. If yields from agriculture per hectare could be improved, this would reduce the need to convert existing forest to farmland to satisfy demand. Farming and the raising of farm animals also produces other greenhouse gases such as methane and oxides of nitrogen. However, a detailed examination of agriculture takes us too far afield, and readers should consult texts and papers referenced by others such as Metz *et al.* (2001). Agricultural practice influences the amount of carbon stored in the ground.

Crop albedo

Different varieties of the same crop have different albedos. Singarayer *et al.* (2009) point out that, by changing the variety of crop, the reflected solar energy can be increased. They increased the reflectance of European crops in a global model by 0.04. Figure 4.2 suggests that the solar energy striking the earth's surface is 55% of the incoming radiation; that is, 188 W m^{-2}. Thus, such a change over the whole earth's surface would increase the reflected energy by 7.5 W m^{-2}. If reflectance change were only over the 10% agricultural area of the earth's surface, the change in radiation forcing would be 0.75 W m^{-2}. For a scenario where the carbon dioxide doubled, they found a cooling of 0.6 °C over Europe.

Soil tilling

The timescale for mobilisation of the carbon below the ground varies widely, from 4 years in the tropics to 1000 years in dry desert conditions. The disruption by tilling the soil to convert grasslands and forests to arable land exposes the soil to the oxygen in the air, resulting in the mobilisation of the carbon and the release of nitrous oxide (N_2O), a strong greenhouse gas. Major sources of nitrous oxide include nitrogen fertilisers and leguminous crop plants. Well-aerated soils also act as a sink of methane.

Livestock grazing

Livestock produce methane in their guts during digestion, which in turn escapes into the environment. When grazing on poor-quality feed, they produce more methane. There has been a steady increase in the number of domestic animals.

Rice paddies

Rice growing is one of the largest agricultural practices in the world, clustered predominantly in Asia, though also in Southern Europe. The process of rice growing requires the flooding of fields, with the rice maturing under this water column. As with any inundation of vegetation (other examples are dams), an anaerobic environment is rapidly created, resulting in decomposition and methane release.

There are a wide variety of agricultural practices that can mitigate carbon dioxide, methane and nitrous oxide emissions. These include:

- The use of crops and crop residues for biofuels to substitute for fossil fuel use, hence reducing fossil fuel-derived CO_2, and providing a new income source for producers.
- Enhancing soil carbon sequestration – a major opportunity for mitigating CO_2.
- Methane capture from manure facilities and its use for fuel; this not only reduces methane emissions but represents a renewable energy source. In Europe and Asia, large livestock operations are installing commercially viable methane

capture and electrical generating facilities. Improvements in technology will likely extend this reach to smaller livestock operations in the future.

- Improving the quality and digestibility of forage. This can reduce methane production from ruminant digestion.
- More efficient use of fertilisers and improved fertiliser technology, which can cut emissions of nitrous oxide as well as improve water quality and cut fertiliser costs for farmers.

Agriculture must be the largest employer in the world, if one considers subsistence farming. There are always alternative practices, which could be explored (e.g. genetically modified grass, which reduces the livestock methane, or genetically modified sheep that don't produce methane). Attempting to change practices that have existed for centuries and generations will be difficult, but influencing large farming cooperatives and governments to adopt best practices in their land use could see the introduction of the *no regrets* options. If changed agricultural practices to reduce greenhouse gas emissions reduced agricultural yield, this implies that more farmland will be needed for the same food output. The gain in greenhouse gas per unit of food may be non-existent.

Biofuel and biochar

Biofuels are being introduced as a way of avoiding an increase in greenhouse gas concentration. Growing the feedstock (such as corn) requires more farmland, and because this is a scarce resource, is driving up the cost of grain. The problem can be envisaged by recognising that a human needs about 9 kJ per day, while a typical car with a 50 kW engine running at half power for 100 minutes a day uses 150 000 kJ per day of energy. A lot of agricultural activity needs to be diverted to fuel production to feed such a hungry entity as the car.

Another possibility is using biochar to store carbon. Plant matter and other organic waste can be subjected to pyrolysis to produce a form of stable carbon that resists biological activity. The pyrolysis process produces gas which can be burned to produce energy. Some of the organic carbon returns to the atmosphere if the flue gases are vented to the atmosphere. However the carbon retained as a solid can be buried during agricultural activity. In some circumstance this is beneficial, but in others biochar can lower soil productivity.

Storing as carbonate

The burning of carbon fuels lowers the energy state of the carbon. Schemes to recover the carbon will require energy input. However, further energy can be obtained when carbon dioxide is reacted with mineral oxides. This method of sequestering carbon by binding it chemically was suggested by Seifritz (1990). Lackner *et al.* (1996) have studied reactions such as those using magnesite in an aboveground process that they claim can proceed with sufficient speed to immobilise the carbon dioxide. Such disposal would be permanent and thermodynamically stable and would not present the risk of accidental release of carbon dioxide of the type possible in geological storage in permeable strata. They imagine the magnesite being obtained by strip mining of surface deposits, and the replacement of magnesium carbonate in the hole made for the mine. They assume that the carbon dioxide is first concentrated such as by capture from power station flue gases. Should the magnesite be moved to the power station, or should the new power stations be built on magnesite mines? Let us consider the latter.

Serpentinite is able to combine with about half its weight of carbon dioxide. This is a similar weight of material to move as in the bicarbonate storage discussed in Chapter 5. If the carbon dioxide came from coal, then the magnesite requires a mine, a number of times larger than the coalmine. The costs come from handling such large amounts of material. An order of magnitude estimate of the cost is as follows.

Capture of CO_2	US$30 per tonne
Digging	US$1 per tonne of serpentinite
Processing	US$5 per tonne of serpentinite
Transport	Variable
Capital	Unknown

Lackner *et al.* (1996) suggest (hope) that in large-scale operations the cost in addition to capture of carbon dioxide could be as low as US$30 per tonne CO_2. Together with capture and transportation costs this would be of the order of US$70. Could the capture cost be avoided by reacting flue gas (15% carbon dioxide) with the magnesium oxides rather than first concentrating the carbon dioxide?

When the same type of process is carried out using the power station cooling water, there is no need for capture and transportation. However the process in Chapter 5 exploits the availability of water. If a dedicated water supply were constructed, the cost would rise considerably.

Enhanced oil recovery

As an oilwell becomes depleted, it is possible to increase its output by injecting carbon dioxide. When the CO_2 mixes with the oil it reduces the oil's viscosity, leading to more flow though the porous strata towards the oilwell; see Figure 6.5. It can also increase the pressure in the oil-bearing strata, which increases the flow

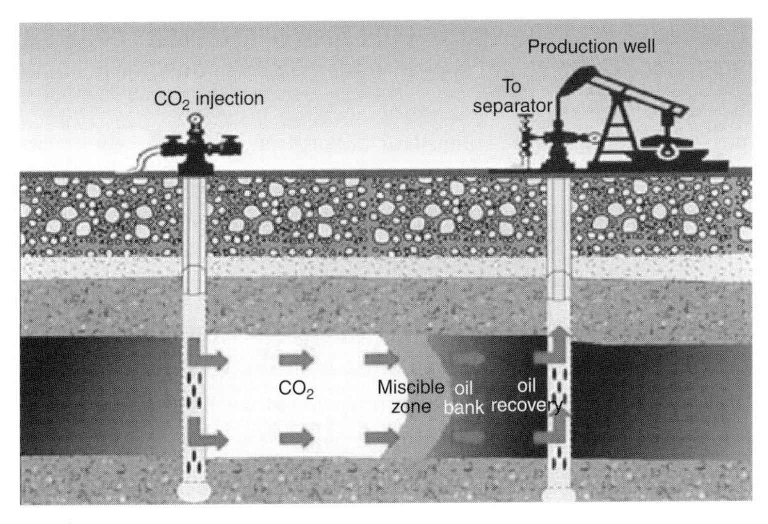

Figure 6.5 See plate section for colour version. Enhanced oil recovery. From Metz *et al.* (2005).

from the oilwell. This process, enhanced oil recovery (EOR) is a well-established industry practice. Stevens *et al.* (2001) estimate that only 5 Mt yr^{-1} of CO_2 that would have been vented to the air is used at present. These sources are mostly ammonia manufacture or CO_2 stripping from natural gas supplies.

This is, however, an industrial use for CO_2 which leads to long-term sequestration. Stevens *et al.* (2001) have estimated from consideration of 155 petroleum provinces that 34 GtC could be sequestered as a *no regrets* option and that the CO_2 could be purchased at US$100 per tonne of carbon.

Carbon dioxide capture

Technologies for separation of carbon dioxide from gas streams have existed for many years and are discussed in many standard chemical engineering texts, for example Maroto-Valer (2010). The original ideas were to dissolve the CO_2 in a solvent and then boil off the CO_2 to obtain a pure stream of CO_2. This process is used in natural gas and ammonia plants.

The order of magnitude cost of separation is US$50 per tonne of CO_2 captured, and as well there is a CO_2 emission penalty. The energy needed to capture a tonne of CO_2 from the flue gas produces CO_2 itself, which means that the net gain is less than one tonne of CO_2. Simmonds and Hurst (2005) discuss the net or 'avoided' cost in terms of dollars per tonne of CO_2 (which accounts for CO_2 emissions associated with the energy demands of the capture process). The cost today is about US$60 per tonne (net). However, there are prospects for lower net costs. Simmonds and Hurst project that the 'best integrated technology' would be able to capture CO_2

(with allowance for the extra emission for the capture energy requirements) for US$28.2 per tonne. These would appear to be reliable figures for post-combustion absorption technologies.

There are four technologies: chemical absorption, pressure swing absorption, membrane technology and cryogenic separation, that can be used to separate carbon dioxide from flue gases. Chemical absorption is often carried out using amines, and they need considerable energy to deabsorb carbon dioxide. In physical absorption, carbon dioxide is dissolved in the absorbant, and the flux of gas into the liquid depends on Henry's Law. The flux is greater for high concentrations of carbon dioxide. The pressure or temperature is then changed to allow the solvent to degas. This is known as swing absorption and needs less energy than amines to achieve separation. Membrane technology allows some molecules to pass through the micropores of the membrane while others are held back. By applying pressure, this process can be accelerated, but at a cost for compression of the flue gases that are mostly nitrogen. One promising approach is to use the membrane at the interface between the flue gas and a liquid absorbant. The sorbent carries away the molecules that pass through the membrane, and this reduces the pressure needed to achieve good rates of separation. Such a system has been tested at a pilot scale in Japan, and it promises a reduction in the energy required.

Rather than attempting to capture the CO_2 after combustion, oxygen can be separated from the atmosphere and used in the place of air for the combustion of coal. The products of combustion then become steam and carbon dioxide. Oxygen separation is usually done cryogenically, a process that exploits the different boiling points of different gases. While the separation of oxygen is expensive, the oxygen cycle (discussed in Chapter 4) has the advantage of greater combustion efficiency.

Land plants and phytoplankton capture carbon directly from their surroundings. Carbon dioxide removal, in the early stages of development by Wright *et al.* (2006), is the concept of capturing carbon dioxide from the ambient air by a base ion exchange resin. Sheets of this material would be exposed to the wind and carbon dioxide captured on the surface of resin sheets. The sheets would then be subjected to high humidity where the resin would give up the carbon dioxide. The released CO_2 would then be compressed and disposed of in geological formations.

Geological sequestration

An extension of the idea of injection of carbon dioxide to enhance oil recovery, is that captured CO_2 can be stored in depleted oil and gas fields or in aquifers or coal seams. There are large natural stores of carbon dioxide underground, and these are used at present to provide high-pressure CO_2. Natural gas is already stored underground, and the capacity at present is 1.6×10^{11} m^3. Many schemes propose sequestration in underground storage and these have been reviewed by Holloway (1997). Some schemes propose underground storage by incorporating the carbon dioxide into basalts.

Depleted oil and gas fields have an immediate attraction, as the geology of the area will be known and the prospect of leakage of the sequestered CO_2 reduced. Freund (2001) estimates depleted oil fields could store 126 $GtCO_2$, and this would more than pay for itself in increased oil production. In the case of gas fields, there is no economic benefit (outside sequestration), and the volume of storage is related to the sequestration cost one is willing to pay. Freund (2001) suggests that 105 $GtCO_2$ could be stored for a cost of less than US\$7 t^{-1} of compressed CO_2, and for higher costs there could be substantially higher amounts.

If carbon dioxide is available as a waste product, this will be an attractive strategy. However capture costs of carbon dioxide are substantial. Goldthorpe and Davidson (2001) suggest for chemical scrubbing or water scrubbing that the cost of capture is of order US\$40 per tonne CO_2. Capital costs are of order US\$75M for 400 MW of flue gas. The units are tonnes of CO_2 avoided.

The costs of compressing the gas to allow injection are also significant. These can be calculated by noting the work done, w, in compressing a volume of gas, dV, from a pressure p_1 to a pressure p_2:

$$w = \int_{p_1}^{p_2} p\mathrm{d}V$$

Typical injection pressures are presented by Gupta *et al.* (2001).

There is much discussion about the leakage of carbon dioxide from underground storage. First there is the dramatic leakage that might occur in an earthquake. Then there is seepage of some fraction of a per cent per year.

Overview of carbon storage

The three strategies for land-based sequestration are injection of carbon dioxide into the ground, chemical reaction with an abundant surface mineral, and enhanced biological fixation of carbon. These are the same three options as for the ocean, and it can be seen that they involve much the same issues. Direct injection (with minor exception) involves both the capture cost and the compression cost. We see that these are substantial. A modest amount of direct injection can benefit oil recovery and so offset the sequestration cost. The procedures involving chemical reaction of the carbon dioxide to form a more benign material involve handing large amounts of mineral, with its subsequent cost. There is no useful product from these chemical reactions and so the expenses of sequestration cannot be offset. The last technique of biological capture is the most interesting. The energy to break the C–O bond is provided 'free' from the sun. The product produced has value – lumber on land or marine protein in the sea. In the case of the land, the forests must compete with agriculture, while in the sea there is, at present, no pressure on the resource. Both have upper limits. The sequestration time on land means forestry is only a holding operation.

The change in the climate as a result of a given change in greenhouse gas concentration in the atmosphere can be estimated by models that embody the laws of physics. They solve the equations that characterise the conservation of heat mass and momentum. The change in economic statistics such as GDP can also be predicted, from numerical models that describe the behaviour of groups of individuals, rather than the laws of physics. There is an important difference, however, between physical models and models of social behaviour. The laws of physics we assume are fixed, while the nature of society is changing. It is also difficult to obtain empirical data on the behaviour of groups of people and so there is a tendency to replace observation with prescription of how one hopes society would behave. A free-market model of the exchange of goods between individuals is a gross simplification of the transactions that actually occur in society. But belief in such abstractions can take on an ideological tone.

Having discussed above the costs of introducing technologies to sequester carbon in the ocean, we would like to use a numerical general equilibrium model of the economy, like that discussed in Appendix 1, to assess the economic benefits of introducing a new energy technology. For a global problem such as climate change one needs a global model. The economic output is calculated based on a number of behavioural assumptions, such as the tradeoff between consumption now and in the future. Some models such as those developed by Nordhaus and Boyer (1999) can divide the economy into a number of like regions to allow for different sociological responses to the same opportunities. For example, the Japanese people are noted for deferring consumption to set aside money for the future, while American people are borrowing now to consume more than they earn.

The Nordhaus models consider the damage caused by climate change as one of the elements of the models. The models prescribe the population in the future and calculate the total GDP. In much the same manner as Chapter 1, the models then estimate the emissions of CO_2. The models contain a simplified (but quite reasonable) climate change calculation which, when used with a curve of economic damage caused by climate change, feeds back into the GDP calculation. If money is spent stopping climate change, the cost is deducted from the investment pool and the climate damage reduced.

The damage function is uncertain, and Nordhaus has used a number of quite different functions. One version is shown in Figure 6.6 as global cost compared with

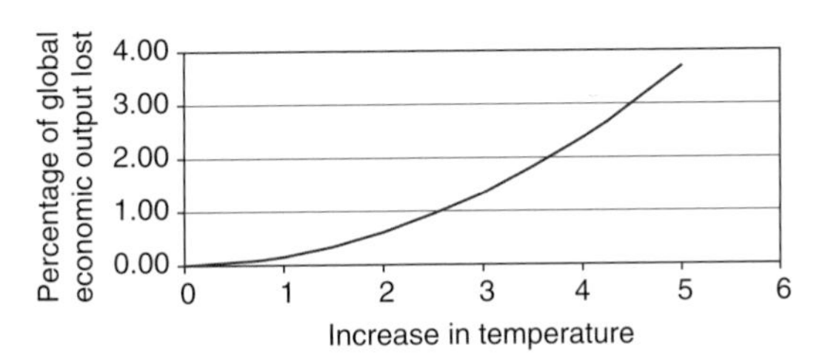

Figure 6.6 Climate damage function, after Nordhaus and Boyer (1999).

the temperature change induced by greenhouse gases. This curve does not allow for a sudden shift in climate. Nor does it recognise that modest carbon dioxide increase accompanied by a temperature rise might provide a benefit in terms of agriculture in cold regions. Even more controversial is the choice of discount factor. A low discount factor models expenditure now to reduce future damage. A high discount factor encourages consumption today and deferral of expenditure on climate change mitigation. In 2006, Sir Nicholas Stern suggested that the present value of climate damage was larger than in most previous studies in part because he assumed a very low discount factor.

Exercise

Exercise 6.1 In Queensland, Australia, it is estimated that 1.05 million hectares of land were cleared, during the two years to August 2003. Calculate the amount of carbon released to the atmosphere and comment on whether the change of albedo would increase or decrease the radiant forcing of the earth.

7

Adaptation

Introduction

The previous chapters looked at how to manage climate change by a number of different approaches. We considered controlling greenhouse gas concentration in the atmosphere or adjusting the solar radiation reflected from the earth back into space. Instead of trying to manage the anthropogenic climate change, the human race could simply adapt to the changes. This option does not seem to have received the attention that it deserves. This is especially true since adaptation is the likely outcome of a lack of resolve to avoid climate change. Lack of resolve comes about from a number of causes. There is uncertainty about the impacts of climate change. There are 'ethical' questions that have been raised about mitigation options. There are people who have decided that reduced consumption is the 'politically correct' behaviour, and harangue others to change their way of living. They oppose schemes to store carbon dioxide and the continued use of fossil fuels. Voluntary reduction of consumption is unlikely to be adopted. There are anti-capitalists who do not want solutions that provide avenues for increased profit. All these diverse opinions inhibit the investment in technology for mitigation.

Climate change can be divided into three categories. First is the slow rate of change that has historically occurred. Changes are small over a human lifetime and are not recognised by the bulk of the people. The human race has adapted to these changes with little problem. Next there is rapid climate change, where changes in climate factors such as rainfall may occur faster than the ability of ordinary innovation to counter. Rapid change might be thought of as the much-discussed rate of temperature change of greater than 2 °C per century. Above this rate of climate change, the routine replacement of infrastructure and migration of people may not be able to cope with the changing environment. Finally there is abrupt climate change, where the changes overwhelm the responses governments can provide, and human disasters eventuate, like the tsunami in the Indian Ocean in 2004, or the earthquake in Haiti in 2010. These, of course, are not the results of climate change but may indicate some of the problems to be faced in adapting to abrupt climate change.

The first strategic response to threatened climate change, as we mentioned above, is to take action to avoid the change; that is, not to do those things which cause climate change, or to do complementary things (such as reducing solar radiation striking the earth) to keep the climate the same. Some potential climate change will occur as a result of 'natural instabilities' in the heat and water balance of the atmosphere and ocean, and we assume humans intervene to avoid these 'natural changes' as well as the human-induced changes. The second strategy is to adapt to the changes in climate. As the temperature fell from a medieval warm period, about 1000 years ago, to the little ice age, people and the environment adapted. For example the Vikings, who had colonised Iceland, Greenland and Newfoundland during the medieval warm period, could not sustain these colonies during colder periods (so it seems) and these settlements withered away.

Adaptation to climate change can be thought of as being in one of two categories. The first is where individuals take action. An example is constructing one's house on stilts to avoid occasional flooding due to rising sea level. This is where one's own resources are used to adapt to the new circumstance. The second category of response is for the community to take action by approaches such as building a levee to hold back the now more-frequent episodes of high water. This division between public expenditure and private expenditures is at the heart of many political differences. It is not economical, or even possible, for an individual person to solve some problems. The 2005 flooding of New Orleans, USA due to inadequate levees is a case of people resisting the taxes necessary to protect the community. Individuals could live elsewhere, but they could do little individually to avoid the problems associated with the flooding of their neighbourhood. This difference in political philosophy has the individualist favouring adaptation to climate change rather than making a present-day expenditure on community causes such as emission reduction schemes.

Rapid climate change

Rapid climate change will be a response to an increasing population, too slow improvements in carbon intensity and a rising consumption (GDP) per person. If the global population were to stop rising and quickly reduce, there would be a reduction in the projected carbon emissions and little climate change to adapt to. Such population changes might be a result of choice or they may be a consequence of rapid climate change. The recent unprecedented rate of population increase brings many problems, and its role in inducing rapid climate change is not always recognised. The increasing population deserves much more attention than it receives.

If carbon intensity were to drop much more rapidly than in the past, there would be less climate change to adapt to. However this will require increased investment in research and development and there is an unwillingness by companies or governments to commit the funds. This is partly due to uncertainty and partly because of low incentive at present. The proponents of 'pricing' carbon dumping say that price would provide the incentive.

Rapid climate change can be avoided by all people consuming less. The question of resource consumption in a finite world is a more general question than just that of climate change. Within living memory all but a small ruling class lived frugally (out of necessity). Thus changing to a less resource-intensive lifestyle in the West (a return to the past) could be considered a better approach than trying to provide, to all, the level of consumption now common in the industrialised countries. This view strikes an emotional chord in many people. The view is driven in part by the truism that most resources are finite. However, we are far from exhausting the most important resources.

As our earlier examination of different scenarios shows, those visions of the future that predict continual economic growth have large emissions of carbon dioxide (if there is no mitigation action). In those scenarios that have increasing greenhouse gas emissions, where all peoples consume at the same rate as the industrialised nations do at present, the practical ability of the earth to sustain this must be questioned. Given the damage being done to the soil, water and atmosphere by the present population, it may not be possible to support extra billions at the level of affluence enjoyed by countries such as in Europe, Australia or Japan. This is not growth limited by the supply of minerals and fossil fuels, but more a limitation of our ability to generate protein. We may not be able to provide, for all, the excess food enjoyed by the developed nations at present. In a situation of lowered consumption, adaption to the climate change may be easily achieved.

Still another possibility is that profligate consumption may be eliminated by a social revolution in attitude about the consumer economy, rather than by necessity. Consumption of fossil fuels may decline and deforestation may be halted. While there are no signs of such a social revolution, if it occurred it would limit greenhouse gas emissions and climate change. Adaptation then would be a minor problem.

The uncertainty about the future climate change is a reason advanced by those that say the world should not make 'economically painful' adjustments until needed. They think we should defer action until the degree of climate change is clearer and then decide whether to adapt or to mitigate. This virtually assures that we will need to adapt to some climate change because the rate that we can arrange mitigation in the future will surely be limited.

What is it that we have to adapt to in a period of climate change? Can we adapt to rising temperature, changing rainfall patterns, rising sea levels, increasing ocean acidity, more frequent extreme events and changing disease prevalence? What will such adaptation cost? Is it more or less than mitigation costs? Let us address the first question.

Possibly the mean temperature of the earth will rise a few degrees. However, we are using temperature as an indicator. As the average temperature varies about 1 °C per three degrees of latitude, the predicted warming of the earth would be like moving a few hundred kilometres towards the equator. In developed countries where people have plenty of disposable income, more air conditioning can easily be installed to keep conditions much as they are now. The impact of this temperature change on human comfort will be small. Even for the poor, the change in temperature may not be such a problem, as already the poor live in a large range of temperatures.

Changing rainfall patterns as a consequence of temperature change is a more challenging problem. Here the agricultural poor are severely disadvantaged. They are in general tied to the land and this is their only capital. If their inherited land cannot be farmed, or cannot produce at historical levels, their options are few.

The sea level may rise. This would require sea walls to be raised and dykes to be installed. What will this cost? Some lands such as the low-lying areas in Bangladesh or some low-lying islands may be abandoned with the consequent large-scale transmigration. Can such shifts of population be achieved without violent conflict?

What is certain is that, as the concentration of carbon dioxide rises in the atmosphere, the amount of carbon invading the upper ocean increases and this drives down the pH. The ocean becomes more acidic. This may lead to significant shifts in the supply of marine protein. Can the coastal states adjust to such a disruption of their food supply?

One of the issues frequently discussed is the possible increase in the number of extreme events. If extreme events became more frequent, the design of many engineering structures would change. For example, offshore structures are often designed to survive the largest wave expected in 100 years. This wave may have a larger magnitude under climate change situations that increase such extreme events.

The changing temperature may directly impact on human health. Rising temperatures will favour some disease carriers such as mosquitoes, but will reduce the number of deaths related to the cold. The modest changes in climate since the Industrial Revolution are credited with causing 140 000 excess deaths per year.

Rapid climate change represents a technical problem for insurance providers, but it is not different in kind from the business they are presently engaged in. Many people think climate change is opposed by insurance companies. Their business relies on there being risk. What they wish is to be able to quantify the risk. Insurance spreads the risk throughout society but it does not reduce the cost of the damage.

Abrupt climate change

By abrupt climate change we mean changes that occur more rapidly than can be accommodated by the established rates of migration, infrastructure replacement or food production methods. For example, the vertical circulation of the oceans consists of the sinking of cold, dense water at the poles, and the general rising in the thermocline in the equatorial regions. This meridional overturning circulation brings the nutrients that rained down from the surface ocean back to the photic zone, supporting the base of the marine food chain. With increased freshwater in the polar regions due to melting ice and the warming of the sea surface, the density of the water that now sinks into the deep ocean may be lowered so much that sinking stops. This would change the supply of the food in the form of fish. This is an example of an event of low probability of occurrence but of important consequence. How will humans adapt to such large changes that are totally outside their experience?

Methane hydrate is a lattice-like structure of water and methane that lies under the polar ocean and the high-latitude permafrost. If disturbed, it can release the methane, which is a powerful greenhouse gas. If large regions of permafrost melted and the methane started to enter the atmosphere, the rate of temperature rise would be accelerated. We might have runaway global warming. How likely is this scenario? Large amounts of methane are stored – twice as much carbon as the other forms of fossil fuel – but not in the permafrost. It is estimated to have a greenhouse gas potential of between 7 and 400 GtC. The methane in marine sediments, on the other hand, is likely to be released slowly if at all. It would seem that the risks of abrupt climate change due to methane are low.

It is hard to see how to adapt to abrupt climate change other than diverting large amounts of resources from activities such as military expenditure into relocating farming and people. The lower the total population, the easier this would seem to be.

Abrupt climate change is a one-off event and is likely to have manifold consequences through large regions of the world. It either happens or it does not; probabilities make no sense. In this case insurance is of little value.

Ethical issues

Humans have adopted various sets of ethical positions. These are a set of principles to which they subscribe. For example, it is considered wrong to kill other humans. For many people their ethical framework is derived from religious beliefs. However, in an increasingly secular world, ethics need another basis and many feel this is difficult to find. Many of the new ethical issues are not directly addressed by religious beliefs.

Changing the global climate as a result of one's actions and making it necessary for others to adapt raises a number of ethical questions. Should the earth be managed for humans or do other species have rights? Do the wealthy have the right to monopolise desired resources and food? Is it ethical to dispose of one's waste in a global commons to the detriment of others?

When making engineering decisions, one should take into account ethical issues. This assertion, however, raises many issues, such as how much weight to accord the varying competing claims on resources and which ethical principles one wishes to follow. Ethical 'principles' are not universally agreed. To overcome this difficulty, ethical considerations are often enshrined in law. Then the might of the State is able to impose these standards on its citizens and to provide retribution on those that transgress. For example, the State declares that it is not legal to steal, and conviction for theft brings a jail sentence.

International agreements on net greenhouse gas emissions are an example of a developing series of laws and penalties to support actions, which we hope are in line with some framework of ethical considerations. What are these considerations? They involve issues such as rich–poor equity, intergenerational equity, precautionary approach to actions, and free trade.

The legal constraints are not enough for some people, and greenhouse gas issues are often presented by such people as an ethical matter. This is in part an effort to evoke responses other than economic self-interest. The US Government in 2002 announced it was not going to take any actions on greenhouse gases that disadvantaged the US economy. To counter such a position, moral or ethical issues are raised in the hope that they will modify the economic rationalist's position. Is it ethical for rich nations to dump the waste products of fossil fuel into the atmosphere thus disadvantaging the agriculture of developing countries? The rich nations whose economies embraced the new source of energy, fossil fuel, 150 years ago, grew rich on the benefits. The dumping of the waste product into the air was the cheapest option. As the atmosphere is a global common, this initially seemed reasonable, as the atmosphere showed little difficulty in adapting to these new inputs of carbon. Now, after 150 years, a large mass of carbon dioxide has built up in the atmosphere. If this continues to increase there is a risk of significant climate change.

The UNFCCC says mitigation actions should not restrict free trade (Grubb, 1999). Why is free trade singled out as a virtue? Some think free trade benefits the strong at the expense of the poor, while others focus on the economic efficiency of free trade. Does the UNFCCC statement have an ethical basis?

When climate change damage involves increased loss of life due to starvation, disease, floods and the like, and life is treated as an economic commodity, new ethical issues emerge. Societal value of life could be measured by either the discounted expected future earnings, the discounted expected future earnings less consumption, or the discounted loss imposed on others due to the death of the person. Especially in societies practising subsistence agriculture, these measures are of little use. They place much greater value in general on the life of a rich, developed-country resident than on the citizen of a LIFD (low-income, food-deficient) country. This is another controversial topic on which we can provide no guidance.

Does the approach of adapting to climate change rather than taking action to stop climate change raise new ethical issues? Before we tackle these questions, we will examine what adaptation to climate change means for food production.

Adaptation and food security

Climate change will endanger food security. *The Green Revolution* on the land has provided the extra food for the rapidly rising world population and, in Asia, has led to a massive adaptation of new agricultural practices. Adaptation was possible. The new technology brings environmental risks, as a monoculture of rice and wheat displaces the more varied species of these cereals previously grown (Huang *et al.*, 2002). Pests are a constant problem of intense farming, requiring continuing management. The consequences have been the overuse of nitrogen, undersupply of other nutrients and too liberal an application of water. Excess water use has led to the salination of soils and the need for more dams with their consequent reduction in river flows. However, despite these consequences of changing 'nature', we are managing most of the undesired consequences by continuous adaptation of agricultural technology. With climate change can we expect the same success?

Overall, the real cost of food has decreased as a result of the Green Revolution, averting the predicted starvation of more millions in the latter part of the twentieth century, but we must recognise that currently there are still 800 million malnourished. The Green Revolution has shown that adjustment can be rapid, although with considerable social costs. Reduced food costs have meant that some marginal farmers were forced off the land. However, once intervention in the ecology has occurred, constant management seems to be needed. Conservative elements in society hope for steady state and are unsettled by this need for constant adaptation. If the change is too large, the society may disintegrate. Thus it is necessary to consider the rate of change. Engineers have been prone in the past to only accounting for direct measurable costs, neglecting or undervaluing the environmental and social costs of their actions. While agriculture has almost kept up, but not caught up with demand, can it continue to do so in the future even with complete mitigation of climate change? We have no answer to this question, and it is subject to the uncertainties about the size of the population and risks of reliance on monocultures.

Without mitigation, the increasing concentrations of carbon dioxide in the atmosphere will in general increase plant growth and reduce their need for water. For crops such as wheat (20% of the world's food), the temperature is an important variable in high-latitude areas such as the Canadian Prairies. Today wheat is grown up to 55 °N, but under doubling of CO_2 in 2050, the wheat-growing area may extend to 65 °N (Ortiz *et al.*, 2008). This would be an adaptation of agricultural practice that would increase food supply. However, increased yield in Canada is of little benefit to the artisan farmer in India, where rainfall has decreased due to climate change.

The danger to food security is in poor countries with many subsistence farmers. If these people need to adapt, they may not be able to due to lack of education and access to capital. If the polluters were assessed for their carbon emissions and the money applied to those disadvantaged, would this be more efficient than spending the money on mitigation? We are not in a position to answer this question.

Intergenerational equity

By dumping greenhouse gases in the atmosphere and relying on adaptation, the present generation is leaving a potential problem for the future generations. Let us assume that climate change that results from the increase in concentration of carbon dioxide disadvantages many people in 50 years time. They have to either adapt or remove the excess carbon dioxide from the atmosphere. Either action is expected to cost money. It can be argued that it is not equitable to leave the problem to future generations. Allowing climate change to occur raises new ethical questions. The present generation is able to consume more today by not reducing the emissions of carbon dioxide. 'They are running up a debt for future generations.' The counter argument is that future generations will be more affluent and can easily cope with the clean up.

Let us think about the present value of this clean up of carbon dioxide. We wish to translate the cost of this future action into equivalent values in today's monetary units. We will use a discount rate of 5% per annum. This is the rate of interest one would earn if the money were deposited in the bank. So if clean up costs US$1 in 50 years time (in dollars that compensate for inflation), its present value is $(1 - 0.05)^{50} = 0.077$ dollars. It makes sense to spend only about seven cents today to avoid a one-dollar problem that will otherwise turn up in 50 years' time. Many people are uncomfortable about this thinking. Goulder and Stavins (2002) address some concerns about the idea of discounting. These include the possibility that if seven cents were put aside, it might not reach those that need to adapt in 50 years. Then there is the difficulty of pricing the restoration of the climate. Providing an extinct species may be very expensive! Trying to save a near-extinct species might prove to be difficult.

However, the funds to aid adaptation could be put aside today. For example, rather than countries that are going to follow the Kyoto protocol spending money to meet the carbon dioxide emission targets, they could put aside an amount of money related to their level of excess emissions over the Kyoto targets. They could put this money in an UNFCCC trust fund to be used to assist in adaptation. How would this money be used? It could be reserved for education. It is easy to believe that, with more education, the presently uneducated would be better able to cope with climate change. Would it be a better strategy to set aside the money to aid adaptation in the future? Maybe the impact of climate change will be mild. Maybe there will be more benefits than negative effects. There is considerable uncertainty. If the benefits do exceed the costs, people might be pleased, then, to have the money for education rather than have 'wasted' it in the early twenty-first century on mitigation.

Case study: Bangladesh

The river delta of the Ganges is a low-lying area already subjected to frequent flooding and loss of life. If the sea level rises as a result of global warming, the area subjected to frequent flooding is expected to increase. Here is a classic case of a group of people who contribute very little to the global warming yet are expected to adapt to the actions of others.

Figure 7.1 illustrates the problem of subsidence and sea-level rise. The one-metre contour of water rise relative to the land would produce a large swathe of land that would be inundated by about 2050 if no action is taken. There are three adaptation strategies available. Build levees, move the people, or accept the more frequent flooding and attempt to provide relief food and accept greater loss of life. The cost of constructing levees will be large and a burden on such a poor country. Possibly a climate change compensation fund might be able to provide the resources. Resettling the people over the next 50 years or so presents difficulties. No country will welcome subsistence farmers as migrants, and resettlement within Bangladesh presents a challenge. The climate change refugees could either be absorbed into the industrial labour force (of which there is a large excess already) or found new land on which to continue the subsistence farming.

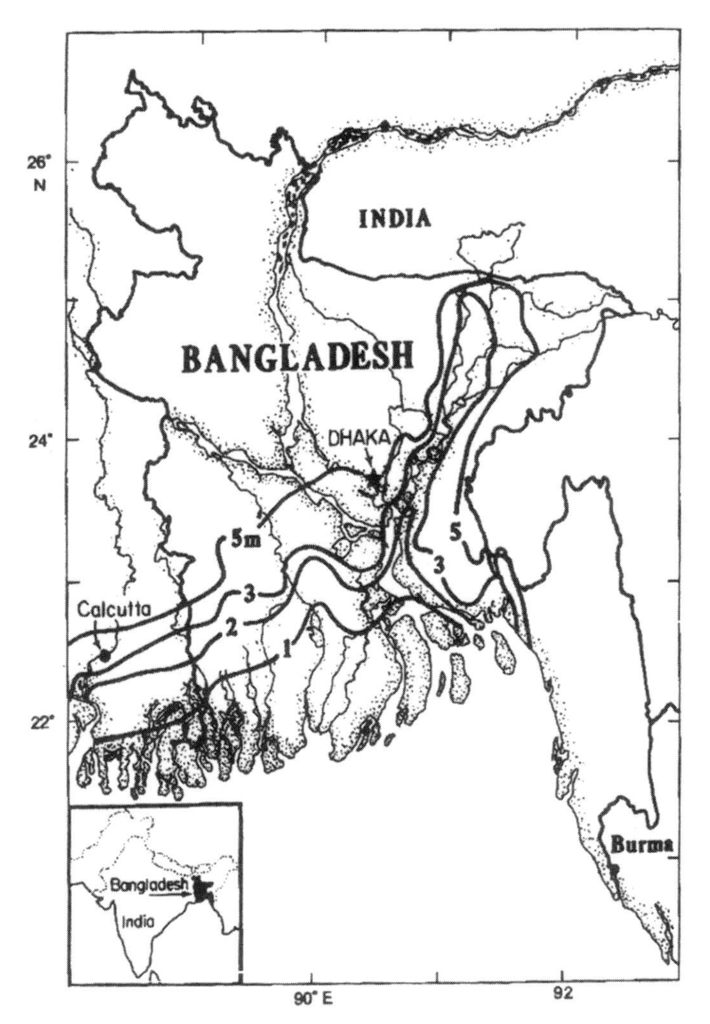

Figure 7.1 Sea level changes will flood low-lying regions of Bangladesh; reproduced from Broadus (1993).

Enhancing adaptive capacity

Predictions of the direction of future climate change will aid adaptation planning. Such predictions need to command some confidence that they are likely to be right more often than they are wrong. Such confidence is hard to create. Short-term predictions of climate change are more likely to be realistic than longer-term predictions. Scientists are aware of the uncertainties that are embedded in present climate models and, while research scientists relish uncertainty, most people are not comfortable with uncertainty. When scientists express legitimate doubts about models of the future climate they lower people's confidence. There are many physical processes not well understood. As well, a climate prediction has a component of sociology. If the model predicts an undesirable change, people may take remedial action to ensure the model predictions do not come true. Thus the model does not predict the future and people lose confidence. Also, human actions that are deemed rational by models, and incorporated into the predictions, may not be followed by humanity. Models of future human behaviour are particularly uncertain. Culture and religion are powerful forces at work, especially amongst those denied education. These forces can lead people not to act in 'their best interest' as seen by outsiders. It has been suggested that education would enhance adaptive capability.

An approach to uncertainty is to make precautionary investments; that is, spend the money to reduce CO_2 emissions, not because it is cost effective, nor an attractive investment in a 'business as usual' scenario, but because it reduces the risk of a catastrophic change in climate. This is 'buying insurance' against the low-probability, high-consequence risk. Alternatively one might invest in research into strains of plants that prosper in a high level of carbon dioxide environment. Precautionary investment can reduce the risks from uncertainties of the future.

The climate has been changing for the last 50 years and people have adapted. The adaptation has mostly been in response to change rather than in anticipation to change. Can we learn from the last 50 years and make adaptation less onerous? Other than improved education to prepare people for change, and improvements in the standard of living to provide the resources to adapt, there is not much one can think of. Improving the standard of living without mitigation of emissions is likely to produce more rapid climate change.

Costs of adaptation

To estimate the costs of adaptation to climate change, one needs to know the potential impacts and their probability of occurrence. We do know that adaptation to climate change by agriculture will involve considerable transition costs. Can we estimate these? Can we estimate the other costs such as changes in the resources per person spent on heating and cooling or in building sea walls to compensate for the changing sea level? These are all market costs. The non-market costs of a changed ecology are even more difficult to quantify. Many of the changes in the environment that we will have to adapt to represent aesthetic costs. In the case of abrupt climate change where food supply is disrupted to the degree that there is mass starvation, it does not seem possible to estimate the cost of climate change in terms of US dollars.

Nordhaus and Boyer (1999) have bravely made such an estimate of the yearly cost of climate change. This is based on an estimate of the greenhouse gas produced and assumes no measures are taken just to reduce greenhouse gas emissions.

(Nordhaus assumes that the carbon intensity, discussed in Chapter 1, changes as we learn to use energy more efficiently.) These estimates are discussed by Jones and Altarawneh (2005), who point out that the costs of mitigation do not consider Ocean Nourishment, the strategy discussed in Chapter 5.

The loss of economic output as a result of changing average surface temperature (and the other environmental factors such as rainfall) has been predicted through climate–economy models. Nordhaus (1994) has used the quadratic expression of Figure 6.6 to relate the loss in the economic output to the increase in the average surface temperature. Nordhaus and Boyer (1999), a few years later, developed a climate damage function that generates a net benefit on a world basis until the predicted increase in the average surface temperature reaches 1.4 °C. This expression is not shown, but Tol (2002) emphasises the uncertainty in such expressions. The damage of climate change to the economy is predicted by an aggregation procedure that adds up the expected damage in different sectors such as agriculture, water resources and human settlements and ecosystems. By choosing a quadratic function of temperature increase, it is assumed that damage from radiative forcing increases smoothly. It assumes there are no sudden large shifts in climate, no surprises. Models include non-market costs (for example, direct impacts on the environment) but generally fail to include social disruptions such as conflict, migration or flight of capital. How much adaptation (in order to reduce damage) there will be in the absence of policy stimulus is uncertain.

Figure 6.6 is an estimate of the fraction of GDP lost if we allow the climate to change. If we spend money to mitigate the climate change, it will also reduce the GDP (except for *no regrets* options, which should not exist in an optimised economic situation). However, some actions will cost less than the reduction in damage. Society will be a little better off doing these rather than doing nothing. The previous chapters have explored the mitigation options and their costs.

One might be surprised that the cost of changing the environment is so low. It is important to remember that humans have already directly changed most of the earth's surface that it is easy to reach. With this changing environment, some species of animals and plants have prospered in certain regions and others suffer a decline. Cattle have prospered in the USA as a consequence of clearing forest for grazing land. Possibly rates of climate change that we presently label as rapid will just cause the present environment to migrate towards or away from the poles as the average temperature of a region changes. How can we assess the difference in value between the environment now and in the future? We could say that any change from the present is undesirable and represents a cost, as Nordhaus has done. This position requires one to say that the present is the most desirable condition. Since we know that the climate has always been changing, it is a difficult position to maintain.

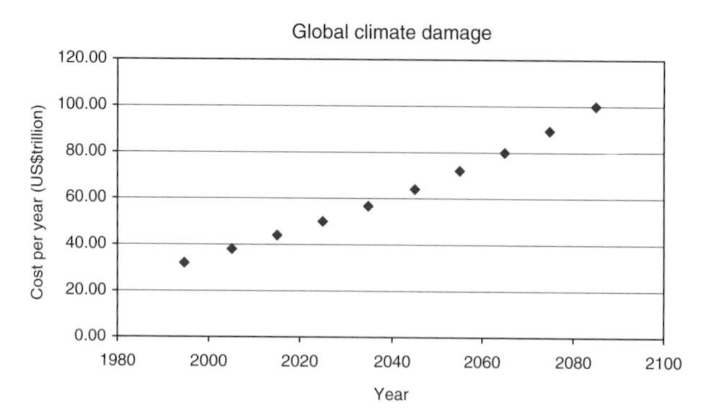

Figure 7.2 The yearly economic cost of climate change, assuming business as usual. The temperature increase by the year 2100 is 2.5 °C. This is a global cost.

This issue about what is a desirable environment involves many personal value judgements. Is forest better than desert? Is it important that the Tasmanian tiger has become extinct? Are these changes just part of evolution? Are there some species we do not wish to become extinct more strongly than others? What is the value, if any, of biodiversity? One anthropocentric view is that we do not want to allow species that might be of value to humans to become extinct. Is that their only value?

The relative importance one places on anthropocentric interests over the aesthetic value of the environments surely depends in part on how far one is from poverty and starvation for oneself and one's children. Since we have 800 million suffering starvation at present, some say the aesthetic value is a luxury we should forgo until all people have enough resources to live in dignity. Until these issues are resolved it is not possible to address the adaptation cost without controversy.

Figure 7.2 contains an estimate of the cost of higher sea walls and changed rainfall patterns. What if you are not sure you need to fix these problems in 20 years time? Say you assess that the probability of needing to fix the problem is only 50%. There is 50% probability that the sea level change is insignificant and that the rainfall patterns stay within the normal ranges. Now on the average one needs to set aside only half the discounted cost today. The low present value of money, and the uncertainty that one needs to fix a problem, encourages most people to choose to take no action. This approach is not appropriate for abrupt climate change when everyone is disadvantaged. There is only one event and it does or does not happen. If it happens, the damage is the whole cost.

Adaptation and equity

Adaptation raises the question of equity. Will adaptation place a bigger burden on some than on others? Undoubtedly! Thus, while the global loss in GDP (Figure

6.6) might only be 1% (for 2.5 °C), some might suffer a much greater loss. Is it ethical to adopt such a course of action – adaptation? Even if the losses from climate change were evenly distributed, is it equitable for the poor, who made small contribution, to be forced to adapt? Is it equitable for those with gains or small losses from climate change to disadvantage others by their actions? To think it is unethical to take actions that are inequitable, one must accept some responsibility for one's fellow human. The free market of Adam Smith did not embrace this concept. It says that by each person applying self-interest the greatest common good is achieved. It is unlikely that Adam Smith was correct.

If people are equal, then each person might be given an equal quota of carbon dioxide emission. If this allowance were the current value of carbon dioxide emission by Americans or Australians, then the concentration of carbon dioxide in the atmosphere would rise quickly. With the sharp rise in GDP of some developing countries such as China, the residents of rapidly developing countries might in the future emit at these levels. Equal net emission quotas per person would need to be just two tonnes of CO_2 per year to stabilise the concentration in the atmosphere. There is an enormous gulf between the different positions of equal emission for each person or emissions frozen at near the situation in 1990.

Some say it is not fair (equitable) that the developed countries limit their output of carbon dioxide unless the developing nations do likewise. This is the position that says the status quo should be preserved much as it is. The concept of using 1990 emissions in the Kyoto Protocol follows this way of thinking. Implicit in this statement is that nations are the important units rather than people. With our focus on sovereign states as a means of governance, the poor are often lumped together within fixed boundaries.

It is reasonable to conclude that adaptation is not equitable unless compensation is part of the concept. It would be made more equitable by transferring resources from the climate beneficiaries to those that are disadvantaged.

Risk assessment

There is both risk assessment and risk management. The first can be imagined as considering the likelihood of your house burning down and the magnitude of the consequences (the US dollar damage). Engineers have the skill to make such an assessment. One way to manage the economic consequences of this risk of fire is to buy fire insurance. This is a judgement based on more information than just the risk of fire. Do you have enough money for food? Is the economic loss unimportant to you? One can see that some risks can be accepted by some but need to be hedged against by others. In the case of climate change, the risks of adverse impacts are much the same for rich nations as for poor (the severity of the consequences are very different). The need to manage the risks is more important for the poor nations, as climate change is likely to bring starvation. We have the unhappy situation that the main generators of the risks are the rich nations (releasing carbon dioxide), while impacts are most damaging for the poor nations.

The precautionary principle was part of the UNFCCC. It says that precautionary measures should be taken despite the absence of full scientific certainty. Engineers of course have always followed the principle of acting without full scientific certainty. As all actions involve risks, the precautionary principle turns out to be a poor guide to decision making.

Many are unhappy with the risks associated with rapid climate change. The acceptability of risks seem to be mostly dependent on the need to take the risk, the degree of control one can exercise and on the fairness of some communities bearing most of the risk but little of the benefits. In the case of CO_2-induced climate change, the developed world receives most of the benefits of fossil fuel burning (most of the coal, oil and gas is consumed by developed countries), while the risks of climate change damage (floods, crop failure, disease increase) will fall on developing countries.

When the change of temperature is only one degree or so, the risks are deemed to be very low for future large-scale discontinuities, but the risk of more extreme weather events will increase. With a temperature change of 5 °C, the risk of a discontinuity in the way of life is higher, while the risk of extreme weather is also much higher. These generalisations about risk are of little value to engineers.

The economies of rich nations are becoming more climate proof as they shift from agriculture to providing services such as banking that are not directly influenced by the climate. This mitigates the risks in the future.

Consequences in Africa and Asia of adapting to climate change

Developing countries are at the greatest risk from climate change. The expected changes in Africa and Asia are reproduced below from the IPCC, together with some measure of the confidence we should attach to the statement.

Africa

- Adaptive capacity of human systems in Africa is low due to lack of economic resources and technology, and vulnerability high as a result of heavy reliance on rain-fed agriculture, frequent droughts and floods and poverty.
- Grain yields are projected to decrease for many scenarios, diminishing food security, particularly in small food-importing countries (*medium to high confidence*).

- Major rivers of Africa are highly sensitive to climate variation; average runoff and water availability would decrease in Mediterranean and southern countries of Africa (*medium confidence*).
- Extension of ranges of infectious disease vectors would adversely affect human health in Africa (*medium confidence*).
- Desertification would be exacerbated by reductions in average annual rainfall, runoff and soil moisture, especially in southern, North, and West Africa (*medium confidence*).
- Increases in droughts, floods and other extreme events would add to stresses on water resources, food security, human health and infrastructures, and would constrain development in Africa (*high confidence*).
- Significant extinctions of plant and animal species are projected and would impact rural livelihoods, tourism and genetic resources (*medium confidence*).
- Coastal settlements in, for example, the Gulf of Guinea, Senegal, Gambia, Egypt and along the east–southern African coast would be adversely impacted by sea-level rise through inundation and coastal erosion (*high confidence*).

Asia

- Adaptive capacity of human systems is low, and vulnerability is high in the developing countries of Asia; the developed countries of Asia are more able to adapt and less vulnerable.
- Extreme events have increased in temperate and tropical Asia, including floods, droughts, forest fires and tropical cyclones (*high confidence*).
- Decreases in agricultural productivity and aquaculture due to thermal and water stress, sea-level rise, floods and droughts, and tropical cyclones would diminish food security in many countries of arid, tropical and temperate Asia; agriculture would expand and increase in productivity in northern areas (*medium confidence*).

Marine protein

The majority of the world's 200 million fishers and their dependents live in areas vulnerable to climate change, according to Allison *et al.* (2005). Some 15% of the animal protein consumed by humans comes from the sea. Away from a narrow coastal fringe, the productivity of the photic zone of the ocean is limited by the supply of nutrients from deeper in the ocean. The thermohaline circulation provides the large-scale recirculation of these nutrients within the ocean basins, while localised upwelling, driven by the surface marine winds, supports regional fisheries. Changes to the ocean temperature in the polar regions, where dense water sinks to drive the thermohaline circulation, and changing wind patterns as a result of global warming, are likely to bring about changes. This possibility is discussed above.

One adaptation strategy to maintain the protein supply from weakening upwelling centres is to provide by other means the nutrients that are no longer available. If the climate changes reduce the supply of nutrients, the concept of Ocean Nourishment discussed in Chapter 5 could be used to restore the productivity of the ocean. Extra nutrients could be manufactured or mined and provided to the ocean to support the base of the food chain. Studies such as that by Ware and Thompson (2005) show a good correlation between phytoplankton standing stock and fish catch. Mesoscale experiments (not in strong upwelling regions) have shown that not only the phytoplankton but also the second trophic level, the zooplankton, respond positively to adding nutrient to the upper ocean.

Exercise

Exercise 7.1 Economic models allow the quantification of costs and benefits of items treated in the money economy. Can you devise a system of attaching value to ethical and environmental issues concerning climate change? Is it a fundamental feature of human nature to try to obtain advantage by dumping waste on communal property?

8

The past and the future

The past

When Svante Arrhenius published his paper in 1896 suggesting that by adding carbon dioxide to the atmosphere the temperature of the earth would be increased, he generated little interest. He used the idea proposed by Joseph Fourier that the earth was warmed by the trapping of heat by the atmosphere, much as in a greenhouse. Arvid Hogbom, a Swede, had pointed out that the amount of carbon dioxide released by humans burning coal was of the same order as the carbon added to the atmosphere from volcanoes and the like.

In scientific circles, the concept of trapping of heat by the addition of carbon dioxide in the atmosphere was not taken seriously until the absorption spectrum was able to be measured with fine enough resolution to show that the carbon dioxide lines did not lie right on top of the water absorption lines.

When Revelle and Suess (1957) calculated, on the prevailing knowledge, that most of the CO_2 released by artificial fuel combustion was absorbed in the ocean, it seemed that there would be no problem. However this was not the case. A molecule of CO_2 left the ocean nearly as often as one entered it, and it was the net flux of carbon dioxide that was important. This problem of stating that there is a large flux of carbon across the sea surface is still bedevilling us, and concepts of net flux needed to be clarified by Ametistova and Jones (2001). The ocean seems to absorb about 2 Gt of carbon per year.

Keeling (1960) started to make accurate measurements in 1958, and by 1960 was able to detect a rise of atmospheric concentration of carbon dioxide. Each year he measured higher concentrations, and his graph of the level of CO_2 began to give an understandable icon of the greenhouse gas effect. After 20 years the problem was considered pressing enough (and still controversial enough) for the World Meteorological Organization and the United Nations Environmental Programme to establish the Intergovernmental Panel on Climate Change (IPCC) in 1988, to review the science behind climate change.

The IPCC has adopted a most interesting approach of trying to establish consensus documents on the science of greenhouse gas. It has issued four assessment reports. These involved hundreds of scientists in the writing and reviewing of the documents. The question of whether the truth is approached more closely by a committee rather than by individuals is open. Much time has been spent debating the process of reaching this consensus. The end result is documents with a non-critical style that quote peer-reviewed literature. The documents tend to quote studies that are not focused on the questions posed. Nor does the IPCC comment on assumptions that are wildly unrealistic.

The first members of the IPCC were appointed in 1988 under the chair of Bert Bolin. In 2007 the IPCC won the Nobel Peace Prize jointly with Al Gore. The second assessment report, *Climate Change 1995* (Houghton *et al.*, 1996) provided much of the underpinning for the Kyoto Protocol of the UNFCCC. This protocol was important as it was an agreement that developed nations would take some actions to reduce their net emissions of greenhouse gas.

Policies for allocating emission rights

In Chapter 5 we introduced the idea of carbon credits and carbon offsets. We used the term *carbon credit* to imply an internationally accepted permit to emit greenhouse gas. Credits are usually issued by governments or the United Nations. It is the legal term and, in the case of hot air trading, need not represent any conscious effort to change emissions. The Russian hot air opportunity due to the political upheaval is such a case. We used the term *offset* as a way of 'offsetting' the increase in greenhouse gas dumped in the atmosphere by undertaking (or having others) conduct an activity that reduces the greenhouse gas in the atmosphere by the equivalent amount.

Both ideas require some concept of a right to emit carbon to the atmosphere. At one extreme, as for CFC emissions, carbon emissions could be banned. The past fossil-fuel emissions that have remained in the atmosphere would decrease initially at a rate of near two to four GtC per year as the ocean and the land take up the carbon. The (reversible) climate change due to fossil fuel burning that has occurred would slowly dissipate. Thus we see there is an element of both amount and time that the carbon remains in the atmosphere. Once there is an emission goal for the world, the question of the allocation of these emissions becomes important.

The emissions allowances could be allocated to land area, or land and EEZ ocean area. They could be allocated on GDP or some other indicator of economic activity. In Chapter 1, it was pointed out that emission could be calculated by taking the country's GDP and multiplying this by the carbon dioxide intensity. This is purely a technique to make future predictions of emissions easier and in no sense implies that allocation of emission 'rights' must or should be allocated on GDP. The final way of allocation is equal per person or equal per person and half for children. These three rules lead to very different allocations. There is no consensus today.

There are a number of catch phrases used in discussions of 'rights'. *Polluter pays*, *willingness to pay*, *comparable burdens* and *existing (historical) rights*. These phrases are not very helpful. Emitting carbon dioxide has an impact at the time and into the future. It is an observation that humans have a time preference, discounting a benefit in the future. This leads to many complicated issues. We showed that increasing the efficiency of fossil fuel use decreased the amount of carbon released to the atmosphere. Economically recoverable fossil fuel is a finite resource. By increasing efficiency today, the fossil fuel saved remains available for future use. We may have only delayed emission, not avoided it, by increasing efficiency. If we store the carbon away from the atmosphere but it slowly leaks back, it has a similar impact in the long term as increasing efficiency. The damage is related to the amount of carbon in the atmosphere and the duration that it remains in the atmosphere. Future damage is discounted by humans and so a tonne of carbon avoided now for 20 years is more valuable (to most people) than half a tonne stored away for 40 years.

An attractive compromise is to allocate emission allowances to political units (countries) and then let the country spend some of the allocation on community activities and distribute the rest to each person. Those in the country that wished to modify behaviour so that they did not use their whole allocation (for example cycling rather than using a motor car) could offset the emissions of those that wished to pay to release more carbon. Some people could change their activities to earn income from their alternative activities. Others could plant trees and sell the offset for so many tonne years of reduced atmospheric reduction. This scheme has advantages over some other schemes as it allows maximum freedom to the individual compared with schemes which aim to penalise those who emit carbon by taxes or arbitrary charges, or changes to uneconomical 'renewable' energy.

There is more on this topic in *Confronting Climate Change* by Mintzer (1992).

Clean Development Mechanism

The Clean Development Mechanism (CDM) is part of the Kyoto Protocol (see Appendix 3) aimed at transferring technology to developing countries. If a developed country undertakes a project in a developing country that reduces the amount of greenhouse gas emitted, either by changing an existing emitting source or by constructing a new facility that is not usual practice, a carbon credit is issued to the developed country. Such carbon credits are traded amongst organisations, some of whom have obligations to hold credits for the carbon dioxide they emit. The CDM process is expected to generate 2.8 $GtCO_2$ certified emission reductions (CERs), by 2012.

The future

Will mitigation be widely adopted? Not if the views of some Americans such as Alan D. Wilson prevail. He asks whether the Kyoto Protocol is not more about money than about concern over global warming. He thinks that Kyoto approval would transfer vast wealth from the United States, Japan and Europe to emerging or poorer nations like China, India, Indonesia and Latin American and African nations. He believes Americans, at a time of military and other important budget needs, would have to sacrifice thousands of US dollars.

It is not at all clear what this transfer of money is, that he is talking about. Is it that Americans will import carbon credits from developing countries? If one believes in free trade, this would be quite legitimate. However, much of the greenhouse gas mitigation proposed under Kyoto will be accomplished by the technologies discussed here. The rich nations can be expected to provide the capital and receive rewards for providing knowledge and expertise. Some of the difficulty is with 'hot air trading', i.e. the transfer of unused permissions to pollute in countries which have undergone an economic downturn since 1990.

Wilson does not consider the damage American carbon dioxide might do to the developing countries by changing the climate. We saw in Chapter 7 that this was of order 300 billion US dollars per year by 2050. The above is a good example of short-term nationalistic thinking on a problem that is global in nature.

Will a more equitable economic system come into being in the future? Will equal emission of greenhouse gas for all people be recognised? Will fossil fuels that could be extracted economically be left in the ground? While a consensus is being formed, the regions of the world that consume much fossil fuel need to implement *no regret* actions and invest research effort in reducing the current uncertainties.

If we wish to stabilise the atmospheric carbon dioxide at 550 ppmv in 100 years time and we assume energy efficiencies are achieved and that there are still many people living in poverty, we will need to sequester some 1000 Gt CO_2 over the next 100 years. This is after we do the easy things. If most of the reduction in emissions comes from the capture of carbon dioxide from flue gases, the cost today (in year 2000 dollars) looks like US$50 per tonne CO_2 (avoided).

Four approaches have been discussed above:

minimum emissions – improved efficiency power plants, cogeneration;
zero-emission technologies. Wind, waves, nuclear energy;
sequestration;
adaptation.

It is the last that people hostile to environmental concerns often favour. In a simple, individual-rights view of the world, concepts such as global commons have little attraction. Letting each person adapt is part of the survival-of-the-fittest view of society.

Exercise

Exercise 8.1 Discuss the history of how it was resolved that the absorption of long-wave radiation in the atmosphere by carbon dioxide was additional to the absorption of water vapour.

Design a scheme for the future that would allocate the rights to dispose of carbon dioxide in the atmosphere.

Appendix 1

Economic costs of CO_2 management

It is generally believed that the reduction of net emissions of carbon dioxide will be achieved more efficiently with tradable carbon credits than without. The subject is treated at length in Freestone (2009) and will not be explored here. The argument is that those organisations that can reduce carbon dioxide emissions or provide carbon sinks more economically than others will sell these benefits to others at a lower cost than the second party could produce the benefit themselves. This is a classic free-market argument.

This is supported by conventional economic discussion. However, the externalities are often neglected or 'wrongly' valued, and many socially undesirable consequences come from applying simple free-market concepts. Engineers in the future will take more account of externalities.

Just as there are climate models that, with the aid of many assumptions, predict the change in the global climate, there are global economic models that try and predict the change in indices such as GDP. The assumptions underlying these models are as uncertain as, or even more uncertain than, in physical models of the atmosphere. We would particularly like to identify the fact that assumptions must be made about social behaviour in the future. Predicting the reaction of people in the future must be considered most challenging, especially when one notes the social changes in large countries such as Russia that have occurred over the last few years.

The simplest economic models have the cost of energy, labour and new investment as inputs, and consumption and investment as outputs. The model must then make an assumption about the impact of perturbations of one of the inputs. They are used in the climate context to examine the impact of reducing energy usage as a function of GDP and switching to more expensive fuels that are less carbon intense.

An alternative model is to create additional sinks and use the lowest cost sum of fuel and sink. Additional sinks can be considered as equivalent to an increase in fuel costs in the same way as fuel switching leads to a higher average price for energy. As we have seen from our examination of tree plantings, there are initially some low cost or even negative cost actions (due to market imperfections) that we designate *no regrets*. As the amount of carbon that it is desired to avoid increases, the cost per tonne of carbon dioxide avoided can be expected to rise. Two sinks of very large capacity that are of almost constant cost, despite the scale, are via geological or ocean direct injection sequestration and Ocean Nourishment. After the already captured CO_2 is sequestered, the price of the first greatly exceeds that of Ocean Nourishment because of the capture cost.

In the economic model by Manne and Richels (1991), they suggest that carbon taxes in excess of US$135 per tonne of carbon (US$34 per tonne CO_2) are needed to achieve worthwhile reductions of CO_2 emission in the USA. At this additional cost for fuel, demand drops off. Manne and Richels suggest losses of US$1.4 trillion at 5% discounting for their base cost of mitigating carbon dioxide emissions. This number sounds large and would seem to us unrealistic. However, it needs to be compared with the total GDP over the last 10 years of US$360 trillion. The amount of mitigation (and so the cost) is sensitive to the assumed future rate of energy efficiency improvement.

If Ocean Nourishment, for example, had been one of the options considered by Manne and Richels, the economic impact would have been much less. One can sense it is important to understand the underlying assumptions in these models. Many are based on 'political philosophy' of the modeller.

Another model is described in Nordhaus and Boyer (2000) and was used in Chapter 7. This DICE model requires the operator to specify the future global population, the change in the productivity of capital and labour, and the fraction of the output consumed rather than invested in capital. Under these assumptions the model predicts the future fossil-fuel consumption. There are many components to this model and they are shown in Figure A.1.

Models that assume that the 'cost' of abatement of CO_2 emissions rises with its magnitude will give very different results to models with sinks that have the capacity to take up large amounts of carbon dioxide at constant cost.

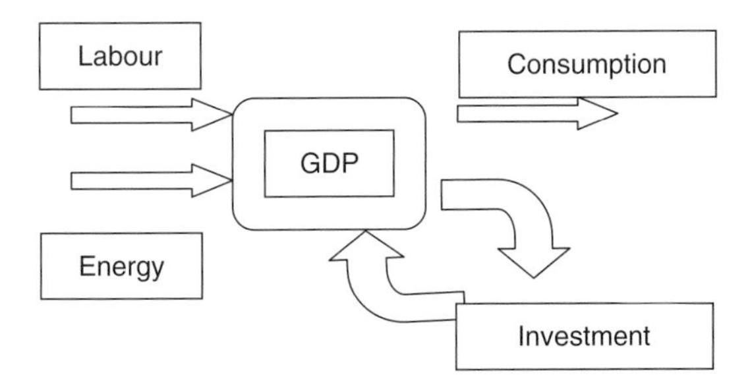

Figure A.1 Traditional economic model.

The above issues present an important dilemma for policy makers. Climate models are uncertain, and more uncertain is the cost of the climate change predicted. Does the world undertake reduction of CO_2 now at considerable economic costs? Are these costs wildly overestimated by economic modellers? If the world delays and then finds climate change creates many problems, it may be too late to reverse the changes at modest cost.

Appendix 2

Present net value and discount rate

It is common when comparing alternative engineering strategies that have a lifetime of several years to calculate the *present net value*. This approach is also known as discounted cash-flow analysis. This approach recognises that income due sometime in the future is not as valuable as income now. The discount rate expresses this as a percentage change per year.

Most projects involve the investment of risk funds (equity), the borrowing of money secured by future income or capital (loan funds), secured by the assets created in the project. Money is expended constructing the plant and then expenses are incurred in operating the plant. Revenue is generated by selling the product produced by the plant.

The concept of present net value recognises that income due sometime in the future is not as valuable as income now. This occurs because income received now can be invested to earn additional income in the future. Thus income is devalued by the discount rate. If the discount rate is $X\%$ pa then US\$100 in one year is worth $100(1 - X/100)$ today. Income of \$1, n years in the future, has a present value of $(1 - X/100)^n$.

A similar approach can be applied to expense. An expenditure of \$1, n years in the future, has a present cost of $(1 - X/100)^n$.

Capital has a cost in terms of the borrowing interest rate. Say this is $Y\%$ per year. Then 100 US dollars borrowed for n years costs $100(Y/100)^n$.

Exercise

Consider this simple example. A product that cost US\$100 in 1990 can be used for 10 years, wherein it has no residual value. It consumes US\$10 per year of materials and generates an income of US\$20 per year. The present value of the project is to be calculated.

Appendix 3

The Kyoto Protocol

The Kyoto Protocol is an international agreement linked to the United Nations Framework Convention on Climate Change. The major feature of the Kyoto Protocol is that it sets binding targets for 37 industrialised countries and the European community for reducing greenhouse gas emissions. These amount to an average of 5% against 1990 levels over the five-year period 2008–2012.

The major distinction between the Protocol and the Convention is that, while the Convention **encouraged** industrialised countries to stabilise greenhouse gas emissions, the Protocol **commits** them to do so.

Recognising that developed countries are principally responsible for the current high levels of greenhouse gas emissions in the atmosphere as a result of more than 150 years of industrial activity, the Protocol places a heavier burden on developed nations under the principle of 'common but differentiated responsibilities'.

The Kyoto Protocol was adopted in Kyoto, Japan, on 11 December 1997, and entered into force on 16 February 2005. To date, 182 Parties of the Convention have ratified its Protocol. The detailed rules for the implementation of the Protocol were adopted at COP 7 in Marrakesh in 2001, and are called the 'Marrakesh Accords'.

The Kyoto mechanisms

Under the Treaty, countries must meet their targets primarily through national measures. However, the Kyoto Protocol offers them an additional means of meeting their targets by way of three international market-based mechanisms.

The Kyoto mechanisms are:

- emissions trading – known as 'the carbon market';
- the Clean Development Mechanism (CDM);
- joint implementation (JI).

The mechanisms help stimulate green investment and help Parties meet their emission targets in a cost-effective way.

Monitoring emission targets

Under the Protocol, countries' actual emissions have to be monitored, and precise records have to be kept of the trades carried out.

Registry systems track and record transactions by Parties under the mechanisms. The UN Climate Change Secretariat, based in Bonn, Germany, keeps an international transaction log to verify that transactions are consistent with the rules of the Protocol.

Reporting is done by Parties by way of submitting annual emission inventories and national reports under the Protocol at regular intervals.

A compliance system ensures that Parties are meeting their commitments and helps them to meet their commitments if they have problems doing so.

Adaptation

The Kyoto Protocol, like the Convention, is also designed to assist countries in adapting to the adverse effects of climate change. It facilitates the development and deployment of techniques that can help increase resilience to the impacts of climate change.

The Adaptation Fund was established to finance adaptation projects and programmes in developing countries that are Parties to the Kyoto Protocol. The Fund is financed mainly with a share of proceeds from CDM project activities.

The road ahead

The Kyoto Protocol is generally seen as an important first step towards a truly global emission reduction regime that will stabilise greenhouse gas emissions, and provides the essential architecture for any future international agreement on climate change.

By the end of the first commitment period of the Kyoto Protocol in 2012, a new international framework needs to have been negotiated and ratified that can deliver the stringent emission reductions that the Intergovernmental Panel on Climate Change (IPCC) has clearly indicated are needed.

Appendix 4

Emission by Annex B countries

Assigned amount, emissions for selected years and demand for and supply of Kyoto units by Annex B Country (1000 tCO_2 equivalent).

Country	Annual assigned amount	Emissions			Kyoto units	
		Base year	2000	2010	2010 demand	2010 supply
Australia	457 925	424 005	500 941	580 000	122 075	
Austria	67 328	77 388	79 754	86 060	18 732	
Belgium	132 034	142 739	152 357	165 300	33 266	
Bulgaria	144 523	157 090	77 697	133 694		10 829
Canada	570 751	607 182	726 249	770 000	199 249	
Croatia	30 348	31 945	22 259	24 500		5 848
Czech Republic	176 657	192 019	147 680	141 700		34 957
Denmark	54 794	69 360	68 505	80 100	25 306	
Estonia	40 014	43 494	19 746	11 660		28 354
Finland	77 093	77 093	73 959	89 900	12 807	
France	559 343	559 343	550 034	688 000	128 657	
Germany	965 984	1 222 765	991 422	979 403	13 419	
Greece	131 119	104 895	130 052	147 206	16 087	
Hungary	95 535	101 633	84 338	62 800		32 735
Iceland	3 078	2 798	2 989	3 494	416	
Ireland	60 679	53 698	66 997	84 656	23 977	
Italy	486 733	520 570	546 905	579 700	92 967	
Japan	1 171 922	1 246 725	1 386 307	1 317 000	145 079	
Latvia	28 570	31 054	11 164	13 000		15 570
Liechtenstein	201	218	218	218	17	
Lithuania	47 425	51 549	34 980	59 148	11 723	
Luxembourg	9 683	13 449	5 971	6 653		3 030
Monaco	93	101	133	127	34	
Netherlands	197 721	210 342	216 916	233 942	36 221	
New Zealand	73 162	73 162	76 955	84 044	10 882	
Norway	52 485	51 965	55 263	63 611	11 126	

Table (*cont.*)

Country	Annual assigned amount	Emissions			Kyoto units	
		Base year	2000	2010	2010 demand	2010 supply
Poland	530 554	564 419	386 187	496 836		33 718
Portugal	82 485	64 949	84 700	82 091		394
Romania	243 689	264 879	204 345	284 368	40 679	
Russia	3 040 332	3 040 332	2 281 100	2 911 800		128 532
Sensitivity				*1 876 690*		*1 163 642*
Slovakia	67 102	72 937	49 165	66 975		127
Slovenia	17 675	19 212	20 697	19 897	2 222	
Spain	329 392	286 428	385 988	343 196	13 804	
Sweden	73 389	70 566	69 357	70 877		2 512
Switzerland	48 975	53 234	52 743	48 190		785
Ukraine	919 220	919 220	454 934	527 314		391 906
United Kingdom	649 681	742 492	649 105	681 626	31 946	
United States	5 701 575	6 130 726	7 001 225	8 115 000	2 413 425	
Total	17 339 267	18 295 967	17 669 337	20 054 085	3 405 316	689 298
Excluding USA and Australia	11 161 767	11 741 236	10 167 171	11 359 085	868 616	689 298

Haites (2004).

Appendix 5

Table of units

Useful numbers

1990 CO_2 emissions of Annex 1	13.63×10^9 t yr^{-1}
1990 NH_4 emissions of Annex 1	1.69×10^9 t yr^{-1} (CO_2 equivalent)
Atmospheric carbon dioxide	750×10^{12} tonnes of carbon
Average atmospheric temperature (2000)	288 K
Barrel of oil	160 litres
Boltzmann constant	5.7×10^{-8} W m^{-2} K^{-4}
CO_2 emissions (1990)	24.2×10^9 t yr^{-1}
Emission/GDP	0.21 kgC US$$^{-1}$ (mixed years)
Emission/GDP	0.77 kgCO_2 US$$^{-1}$ (mixed years)
Emissions (1998)	6.608×10^9 tC yr^{-1} (fossil fuel burning and cement manufacture)
Installed electric power generation 2000	3300 GW
GNI per capita (2002)	US$5170
Molecular weight of C	12
Molecular weight of CO_2	44
Partial pressure of CO_2 (2000)	370 ppmv
Population (2002)	6.1×10^9
Population growth (1980–1999)	1.6% pa (average)
Radius of the earth	6356 to 6378 km
Solar 'constant'	1367 W m^{-2}
Surface area of ocean	3×10^{14} m^2
Surface area of the land	148×10^{12} m^2
Surface area of the world	510×10^{12} m^2
Volume of the ocean	1.370×10^9 km^3
World GDP (2002)	US$$31.5 \times 10^{12}$
1 watt for a year	31.5 MWh
kilo	10^3
mega	10^6
giga	10^9
tera	10^{12}
peta	10^{15}
exa	10^{18}

CE = current era

W = watts

t = tonnes = 1000 kg

CCOS = carbon capture and ocean storage

LIFD = low-income, food-deficient

UNFCCC = United Nations Framework Convention on Climate Change

Note: US$ are year 2002 value.

Natural gas methane conversion factors.

Unit	Multiplication factor	Unit
Cubic foot (ft^3) CH_4	0.042 46	Pound (lb) CH_4
Cubic foot (ft^3) CH_4	1 014.6[a]	Btu
Cubic foot (ft^3) CH_4	0.2974	Kilowatt-hour (kWh)
Cubic foot (ft^3) CH_4	0.000 404	Tonne CO_2 equivalent
Cubic foot (ft^3) CH_4	0.000 11	Tonne C equivalent
Cubic foot (ft^3) CH_4	19.26	Gram (g)
Pound (lb)[b] CH_4	23 896	Btu
Pound (lb) CH_4	7	Kilowatt-hour (kWh)
Pound (lb) CH_4	0.009 53	Tonne C equivalent
Pound (lb) CH_4	0.0026	Tonne C equivalent
Btu	0.000 293	Kilowatt-hour (kWh)
Kilowatt-hour (kWh)	3 600 000	Joule
Btu	1.054	Kilojoule
CH_4 GWP[c]	21	CO_2 GWP
CO_2	0.272 73	C equivalent
CH_4	5.7273	C equivalent

[a] Higher heating value (HHV) @ 60 °F. 30″ Hg. dry.

[b] One pound = 0.45 kg.

[c] GWP = global warming potential over a 100-year time frame (GWP for carbon dioxide = 1).

Appendix 6

Inflation table

International Monetary Fund (IMF) inflation table: change per year (%).

Year	Advanced economies	World
2000	2.3	4.6
2001	2.2	4.3
2002	1.6	3.5
2003	1.9	3.7
2004	2.0	3.6
2005	2.3	3.8
2006	2.4	3.7
2007	2.2	4.0
2008	3.4	6.0
2009	0.1	2.5
2010	1.4	3.7

www.imf.org/external/datamapper/index.php?chart=barchartView&db=WEO.

Further reading

The following is a short list of books that provide further background reading for the student. Note both old and new ISBN are provided.

1. *The Oceans and Climate* (2nd Edition)
   Grant R. Bigg
   (ISBN: 0521016347 | ISBN-13:9780521016346)
   DOI: 10.2277/0521016347 551.46 BIG
2. *The Science and Politics of Global Climate Change*
   Andrew E. Dessler, Edward A. Parson
   (ISBN: 0521831709 | ISBN-13:9780521831703)
   DOI: 10.2277/0521831709
3. *Climate Variability, Climate Change and Fisheries*
   Edited by Michael H. Glantz
   (ISBN: 0521017823 | ISBN-13:9780521017824)
   DOI: 10.2277/0521017823
4. *Emissions Trading for Climate Policy*
   Edited by Bernd Hansjürgens
   (ISBN: 0521848725 | ISBN-13:9780521848725)
   DOI: 10.2277/0521848725
5. *Carbon Dioxide Capture and Storage*
   IPCC
   (ISBN: 0521685516 | ISBN-13:9780521685511)
   DOI: 10.2277/0521685516 (chapter on oceans)
6. *Confronting Climate Change*
   Edited by Irving M. Mintzer
   (ISBN: 0521421098 | ISBN-13:9780521421096)
   DOI: 10.2277/0521421098
7. *Population and Climate Change*
   Brian C. O'Neill, F. Landis MacKellar, Wolfgang Lutz
   (ISBN: 0521018021 | ISBN-13:9780521018029)
   DOI: 10.2277/0521018021
8. *Avoiding Dangerous Climate Change*
   Edited by Hans Joachim Schellnhuber, Wolfgang Cramer, Nebojsa Nakicenovic, Tom Wigley, Gary Yohe; Foreword by Tony Blair; Introduction by Rajendra Pachauri

(ISBN: 0521864712 | ISBN-13:9780521864718)
DOI: 10.2277/0521864712

9. *Climate Change 2001: Synthesis Report*
   Edited by Robert T. Watson
   (ISBN: 0521015073 | ISBN-13:9780521015073)
   DOI: 10.2277/0521015073
10. *Climate Change Policy*
   Catrinus J. Jepma, Mohan Munasinghe; Foreword by Bert Bolin, Robert Watson, James P. Bruce
   (ISBN: 052159314X | ISBN-13:9780521593144)
   DOI: 10.2277/052159314X
11. *The Economics of Climate Change. The Stern Review*
   Nicholas Stern
   (ISBN: 0521700809| ISBN-13:9780521700801)
   DOI: 10.2277/0521700809
12. *The Skeptical Environmentalist*
   Bjørn Lomborg
   (ISBN: 0521010683 | ISBN-13: 9780521010689)
   DOI: 10.2277/0521010683

References

Adams, E.E., Golomb, D., Zhang, X.Y. and Herzog, H.J. (1995) Confined release of CO_2 into shallow seawater, in N. Handa and T. Ohsumi (eds.), *Direct Ocean Disposal of Carbon Dioxide*, Terrapub, pp. 153–164.

Alcamo, J., Kreileman, G.J.J., Krol, M.S. and Zuidema, G. (1994a) Modelling the global society-biosphere-climate system. Part 1: model description and testing. *Water, Air and Soil Pollution*, **76**, 1–35.

Alcamo, J., van den Born, G.J., Bouwman, A.F. *et al.* (1994b) Modelling the global society-biosphere-climate system. Part 2: computed scenarios. *Water, Air and Soil Pollution*, **76**, 37–78.

Allison, E.H., Adger, W.N., Badjeck, M.C. *et al.* (2005) Effects of climate change on the sustainability of capture and enhancement fisheries important to the poor: analysis of the vulnerability and adaptability of fisherfolk living in poverty. Fisheries Management Science Programme London, Department for International Development Final Technical Report. Project No. R4778J. Available online at www.fmsp.org.uk.

Ametistova, L. and Jones, I.S.F. (2001) *Revising the Understanding of Carbon Dioxide Fluxes into the Sea*, IOC/WESTPAC.

Ametistova, L., Briden, J. and Twidell, J. (2002) The sequestration switch: removing industrial CO_2 by direct ocean absorption. *The Science of the Total Environment*, **289**, 213–223.

Anderson, R., Brandt, H., Mueggenburg, H., Taylor, J. and Viteri, F. (1999) A power plant concept which minimizes the cost of carbon dioxide sequestration and eliminates the emission of atmospheric pollutants, in P. Riemer, B. Eliasson and A. Wokaun (eds.), *Proceedings of the Fourth International Conference on Carbon Dioxide Removal, Interlaken, Switzerland, Aug. 30 – Sept. 2, 1998*, Pergamon, pp. 59–61.

Anonymous (1963) Review of research and development in cloud physics and weather modification. *Soviet-Block Research in Geophysics, Astronomy and Space*, **83**, 24512.

Arrhenius, S. (1896) On the influence of carbonic acid in the air upon the temperature on the ground. *The London, Edinburgh and Dublin Magazine and Journal of Science, 5th series*, 237–276.

Auerbach, D.I., Caufield, J.A., Adams, E.E. and Herzog, H.J. (1997) Impacts of ocean CO_2 disposal on marine life: I. a toxicological assessment integrating constant-concentration laboratory assay data with variable-concentration field exposure. *Environmental Modeling and Assessment*, **2**, 333–343.

Aumont, O. and Bopp, L. (2006) Globalizing results from ocean in situ iron fertilization studies. *Global Biogeochemical Cycles*. **20**, GB2017. DOI: 10.1029/2005GB002591.
Aya, I., Yamane, K. and Nariai, H. (1997) Solubility of CO_2 and density of CO_2 hydrate at 30 MPa. *Energy*, **22**, 263–271.
Aya, I., Yamane, K. and Shiozaki, K. (1999) Proposal of self sinking CO_2 sending system: COSMOS, in P. Riemer, B. Eliasson and A. Wokaun (eds.), *Greenhouse Gas Control Technologies*, Elsevier, pp. 269–274.
Baes, C.F., Bjorkstrom, A. and Mulholland, P.J. (1985) Uptake of carbon dioxide by the oceans, in J.R. Trabalka (ed.), *Atmospheric Carbon Dioxide and the Global Carbon Cycle. DOE/ER-0239*, United States Department of Energy, pp. 81–111.
Balch, W., Evans, R., Brown, J. *et al.* (1992) The remote sensing of ocean primary productivity: use of a new data compilation to test satellite algorithms. *Journal of Geophysical Research*, **97**, 2279–2293.
Bekkeheien, M. (1995) *Global Transport Sector Energy Demand towards 2020*, World Energy Council.
Bilger, R.W. (1999) Zero release combustion technologies and the oxygen economy, in *Proceedings of the Fifth International Conference on Technologies and Combustion for a Clean Environment, Lisbon, Portugal 12–15 July, 1999*, pp. 1039–1046.
Birdsall, N. (1994) Another look at population and global warming, in *Population Environment and Development: Proceedings of the United Nations Expert Group Meeting on Population, Environment and Development, New York City, 20–24 January 1992*. ST/ESA/SER.R/129, United Nations, pp. 39–52.
Brewer, P.G. (2000) Contemplating action: storing carbon dioxide in the ocean. *Oceanography*, **13**, 84–92.
Brewer, P.G., Goyet, C. and Friederich, G. (1997) Direct observation of the oceanic CO_2 increase revisited. *Proceedings of the National Academy of Sciences of the United States of America*, **94**, 8308–8313.
Broadus, J.M. (1993) Possible impacts of, and adjustments to, sea-level rise: the case of Bangladesh and Egypt, in R.A. Warrick, E.M. Barrow and T.M.L. Wigley (eds.), *Climate and Sea Level Change: Observations, Predictions and Implications*, Cambridge University Press, pp. 263–275.
Buesseler, K.O. and Boyd, P.W. (2003) Will ocean fertilisation work? *Nature*, **300**, 67–68.
Buitenhuis, E.T., van der Wal, P. and de Baarl, H.J.W. (2001) Blooms of Emiliania huzleyi are sinks of atmospheric carbon dioxide: a field and mesocosm study derived simulation. *Global Biogeochemical Cycles*, **15**, 577–587.
Caldeira, K. (2003) Simulating ocean fertilization: effectiveness and unintended consequences, in the National Academy of Engineering, *The Carbon Dioxide Dilemma: Promising Technologies and Policies*, National Academies Press, pp. 53–58.
Caldeira, K. and Rau, G.H. (2000) Accelerating carbon dissolution to sequester carbon dioxide in the ocean: geochemical implications. *Geophysical Research Letters*, **27**, 225–228.
Caldeira, K. and Wickett, M.E. (2003) Anthropogenic carbon and ocean pH. *Nature*, **425**, 365.
Coale, K.H., Johnson, K.S., Fitzwater, S.E. *et al.* (1996) A massive phytoplankton bloom induced by an ecosystem-scale iron fertilization experiment in the equatorial Pacific Ocean. *Nature*, **383**, 495–501.
Coale, K.H., Johnson, K.S., Chavez, F.P. *et al.* (2004) Southern Ocean iron enrichment experiment: carbon cycling in high- and low-Si waters. *Science*, **304**, 408–414. DOI: 10.1126/science.1089778.

Crutzen, P.J. (2006) Albedo enhancement by stratospheric sulphur injections; a contribution to resolve a policy dilemma? *Climatic Change*, **77**, 211–219. DOI: 10.1007/s10584–006–9101-y.

Davidson, E.A. and Hirsch, A.I. (2001) Fertile forest experiments. *Nature*, **411**, 431–433.

Delille, B., Harlay, J., Zondervan, I. *et al.* (2005) Response of primary production and calcification to changes in pCO_2 during experimental blooms of the coccolithophorid Emilliania huxleyi. *Global Biogeochemical Cycles*, **19**, GB2023. DOI: 10.1029/2004GB002318.

Edington, A.J. (2005) *Ocean nourishment in the high seas*, ME Thesis, University of Sydney, Australia.

Falkowski, P.G. and Raven, J.A. (1997) *Aquatic Photosynthesis*, Blackwell Malden.

Fligge, M. and Solanki, S.K. (2000) The solar spectral irradiance since 1700. *Geophysical Research Letters*, **27**, 2157–2160.

Freestone, D. (2009) *Legal Aspects of Carbon Trading: Kyoto, Copenhagen, and Beyond*, Oxford University Press. ISBN: 9780199565931.

Freund, P. (2001) Progress in understanding the potential role of CO_2 storage, in D.J. Williams, R.A. Durie, P. McMullan, C.A.J. Paulson and A.Y. Smith (eds.), *Proceedings of the Fifth International Conference on Greenhouse Gas Control Technologies*, CSIRO Publishing, pp. 272–277.

Gnanadesikan, A., Sarmiento, L.J. and Slater, R.D. (2003) Effects of patchy ocean fertilization on atmospheric carbon dioxide and biological production. *Global Biogeochemical Cycles*, **17**, 1050. DOI: 10.1029/2002GB001940.

Goff, F. and Lackner, K.S. (1998) Carbon dioxide sequestering using ultramafic rocks. *Environmental Geosciences*, **5**, 89–101.

Goldthorpe, S. and Davidson, J. (2001) Capture of CO_2 using water scrubbing, in D.J. Williams, R.A. Durie, P. McMullan, C.A.J. Paulson and A.Y. Smith (eds.), *Proceedings of the Fifth International Conference on Greenhouse Gas Control Technologies*, CSIRO Publishing, pp. 155–160.

Golomb, D. and Angelopoulos, A. (2001) *A Benign Form of CO_2 Sequestration in the Ocean. First National Conference on Carbon Sequestration*, NETL publications.

Goulder, L.H. and Stavins, R.N. (2002) An eye on the future: how economists' controversial practice of discounting really affects the evaluation of environmental policies. *Nature*, **419**, 673–674.

Grubb, M. (1999) *The Kyoto Protocol: A Guide and Assessment*. The Royal Institute of International Affairs, p. 39. ISBN 1 85383 581 1.

Gupta, N., Wang, P., Sass, B., Bergman, P. and Byrer, C. (2001) Regional and site-specific hydrogeological constraints on CO_2 sequestration in the Midwestern United States saline formations, in D.J. Williams, R.A. Durie, P. McMullan, C.A.J. Paulson and A.Y. Smith (eds.), *Proceedings of the Fifth International Conference on Greenhouse Gas Control Technologies*, CSIRO Publishing, 385–390.

Haites, E. (2004) Estimating the market potential for the clean development mechanism: review of models and lessons learned. PCF*plus* Report 19, World Bank, Washington.

Haugen, H.A. and Eide, L.I. (1996) CO_2 capture and disposal: the realism of large scale scenarios. *Energy Conversion and Management*, **37**, 1061–1066.

Hertzog, H.G. (1999) Ocean sequestration of CO_2 – an overview, in P. Riemer, B. Eliasson and A. Wokaun (eds.), *Greenhouse Gas Control Technologies*, Elsevier, pp. 237–242.

Hirai, S., Okazaki, K., Tabe, Y. and Hijikata, K. (1997) Mass transport phenomena of liquid CO_2 with hydrate. *Waste Management*, **17**, 353–360.

Hoffert, M., Wey, Y.-C., Callegari, A.J. and Broecker, W.S. (1979) Atmospheric response to deep-sea injections of fossil-fuel carbon dioxide. *Climate Change*, **2**, 53–68.

Holloway, S. (1997) An overview of the underground disposal of carbon dioxide. *Energy Conversion and Management*, **38S**, 193–198.

Houghton, J.T., Jenkins, G.J. and Ephraums, J.J. (1990) *Climate Change: The IPCC Scientific Assessment*, Cambridge University Press.

Houghton, J.T., Callander, B.A. and Varney, S.K. (1992) *Climate Change 1992*, Cambridge University Press.

Houghton, J.T., Meira Filho, L.J., Callander, B.A. *et al.* (1996) *Climate Change 1995: The Science of Climate Change*, Cambridge University Press.

Houghton, J.T., Ding, Y., Griggs, D.J. *et al.* (eds.) (2001) *Climate Change 2001: The Scientific Basis. Contribution of Working Group I to the Third Assessment Report of the IPCC*, Cambridge University Press.

Howarth, R.W. (1988) Nutrient limitation of net primary production in marine ecosystems. *Annual Review of Ecology and Systematics*, **19**, 89–110.

Huang, J., Pray, C. and Rozelle, S. (2002) Enhancing the crops to feed the poor. *Nature*, **418**, 678–684.

Jamieson, D. (1996) Ethics and intentional climate change. *Climatic Change*, **33**, 323–336.

Jin, X. and Gruber, N. (2003) Offsetting the radiative benefit of ocean iron fertilization by enhancing N_2O emissions. *Geophysical Research Letters*, **30**, 2249. DOI: 10.1029/2003GL018458.

Johnston, P., Santillo, D., Stringer, R. *et al.* (1999) *Ocean Disposal/Sequestration of Carbon Dioxide from Fossil Fuel Production and Use: an Overview of Rationale, Techniques and Implications. Greenpeace Research Laboratories, Technical Note 01/99*, Greenpeace International.

Jones, I.S.F. (2004) The enhancement of marine productivity for climate stabilization and food security, in A. Richmond (ed.), *Handbook of Microalgal Cultures*, Blackwell, Chapter 33, pp. 534–545.

Jones, I.S.F. and Altarawneh, M. (2005) The economics of CO_2 sequestration scenarios using ocean nourishment, in *Conference Proceedings: Fourth Annual Conference on Carbon Capture and Sequestration DOE/NETL, 2–5 May, 2005, Alexandria, Virginia, USA*. http://hdl.handle.net/2123/1422.

Jones, I.S.F. and Caldeira, K. (2003) Long-term ocean carbon sequestration with macronutrient addition, in *Proceedings of the Second Annual Conference on Carbon Sequestration: Developing & Validating the Technology Base to Reduce Carbon Intensity, Alexandria, VA, May 5–8, 2003*. www.netl.doe.gov/publications/proceedings/03/carbon-seq/PDFs/110.pdf.

Jones, I.S.F. and Cappelen-Smith, C. (1999) Lowering the cost of carbon sequestration by ocean nourishment, in P. Riemer, B. Eliasson and A. Wokaun (eds.), *Greenhouse Gas Control Technologies*, Elsevier, pp. 255–259.

Jones, I.S.F. and Otaegui, D. (1997) Photosynthetic greenhouse gas mitigation by ocean nourishment. *Energy Conversion and Management*, **38S**, 379–384.

Jones, I.S.F. and Young, H.E. (1997) Engineering a large sustainable world fishery. *Environmental Conservation*, **24**, 99–104.

Jones, I.S.F., Sugimori, Y. and Stewart, R.W. (1993) *Satellite Remote Sensing of the Oceanic Environment*, Seibutsu Kenkyusha.

Judd, B.J., Harrison, D.P. and Jones, I.S.F. (2008) *Engineering Ocean Nourishment*, World Conference of Engineering. http://hdl.handle.net/2123/2664.

Keeling, C.D. (1960) The concentration and isotopic abundances of carbon dioxide in the atmosphere, *Tellus*, **12**, 200–203.

Keith, D.W. (2000) Geoengineering the climate: history and prospect. *Annual Review of Energy and the Environment*, **25**, 245–284.

Keith, D.W. and Dowlatabadi, H. (1992) A serious look at geoengineering. *EOS, Transations American Geophysical Union*, **73**, 289–293.

Kosugi, S., Hinada, Y., Watanabe, K. *et al.* (1999) A preliminary study of the gas-lift advanced dissolution (GLAD) system in deep sea sequestration of CO_2, in *Proceedings of OMAE'99, 18th International Conference on Offshore Mechanics and Arctic Engineering, July 11–16, 1999, St John's Newfoundland, Canada, OMAE99/OSU-3051.*

Lackner, K.S., Butt, D.P., Wendt, C.H. and Sharp, D.H. (1996) Carbon dioxide disposal in solid form, LA-UR-96–598, in B.A. Sakkestad (ed.), *Proceedings of the 21st International Technical Conference on Coal Utilization and Fuel Systems*, Coal and Slurry Technology Association, pp. 133–144.

Langdon, C., Takahashi, T., Sweeney, C. *et al.* (2000) Effect of calcium carbonate saturation state on the calcification rate of an experimental coral reef. *Global Biogeochemical Cycles*, **14**, 639.

Learn, J., Beer, J. and Bradley, R. (1995) Reconstruction of solar irradiance since 1610: implications for climate change. *Geophysical Research Letters*, **22**, 3195–3198.

Levitus, S., Boyer, T.P. and Antonov, J. (1994) World Ocean Atlas 1994, Volume 5. Interannual variability of upper ocean thermal structure. Technical Report PB–95–270120/XAB;NOAA-ALTAS–5, National Environmental Satellite, Data, and Information Service, Washington, DC, USA.

Liss, P.S. and Merlivat, L. (1986) Air–sea gas exchange rates: introduction and synthesis, in P. Buat-Ménard (ed.), *The Role of Air–Sea Exchange in Geochemical Cycling*, Reidel, pp. 113–127.

Loh, H.P., Ruether, J. and Dye, R. (1998) Advanced technologies for power generation with reduced carbon dioxide emissions, in R.W.F. Riemer, A.Y. Smith and K.V. Thambimuthu (eds.), *Greenhouse Gas Mitigation*. Elsevier, pp. 567–572.

Lovelock, J.E. and Rapley, C.G. (2007) Ocean pipes could help the Earth to cure itself. *Nature*, **449**, 403.

Manne, A.S. and Richels, R.G. (1991) Global CO_2 emission reductions – the impacts of rising energy costs. *The Energy Journal*, **12**, 87–108.

Marchetti, C. (1977) On geoengineering and the CO_2 problem, *Climatic Change*, **1**, 59–68.

Marinov, I., Follows, M., Gnanadesikan, A., Sarmiento, J.L. and Slater, R.D. (2008) How does ocean biology affect atmospheric pCO2? Theory and models. *Journal of Geophysical Research*, **113**, C07032. DOI: 10.1029/2007JC004598.

Markels, M. and Barber, R.T. (2001) Sequestration of CO_2 by ocean fertilization, Poster Presentation for NETL 1st National Conference on Carbon Sequestration, May 14–17, 2001, Washington, DC. www.netl.doe.gov/publications/proceedings/01/carbon_seq/p25.pdf.

Maroto-Valer, M.M. (2010) *Developments and Innovation in Carbon Dioxide (CO_2) Capture and Storage Technology, Volume 1: Carbon Dioxide (CO_2) Capture, Transport and Industrial Applications*, CRC Press.

Martin, J.H., Fitzwater, S.E. and Gordon, R.M. (1990) Iron deficiency limits phytoplankton growth in Antarctic waters. *Global Biogeochemical Cycles*, **4**, 5–12.

Martin, J.H., Coale, K.H., Johnson, K.S. *et al.* (1994) Testing the iron hypothesis in ecosystem of the equatorial Pacific Ocean, *Nature*, **371**, 123–124.

Marubini, F. and Thake, B. (1999) Bicarbonate addition promotes coral growth. *Limnology and Oceanography*, **44**, 716–720.

Matear, R.J. and Elliot, B. (2004) Enhancement of oceanic uptake of anthropogenic CO_2 by macro-nutrient fertilization. *Journal of Geophysical Research*, **109**. DOI: 10.1029/2000JC000321.

Metz, B., Davidson, O., Swart, R. and Pan J. (eds.) (2001) *Climate Change 2001: Mitigation Contribution of Working Group III to the Third Assessment Report of the Intergovernmental Panel on Climate Change (IPCC)*, Cambridge University Press.

Metz, B., Davidson, O., de Coninck, H., Loos, M. and Meyer, L. (eds.) (2005) *IPCC Special Report on Carbon Dioxide Capture and Storage*. Cambridge University Press.

Mintzer, I.M. (ed.) (1992) *Confronting Climate Change. Risks, Implications and Responses*, Cambridge University Press.

Moore, J.K., Doney, S.C., Lindsay, K., Mahowald, N. and Michaels, A.F. (2006) Nitrogen fixation amplifies the ocean biogeochemical response to decadal timescale variations in mineral dust deposition. *Tellus*, **58B**, 560–572.

Najjar, R.G. and Keeling, R.F. (2000) Mean annual cycle of the air–sea oxygen flux: a global view. *Global Biogeochemical Cycles*, **14**, 573–584.

Nakashiki, N., Ohsumi, T. and Katano, N. (1995) Technical view on CO_2 transportation onto the deep ocean floor and dispersion at intermediate depths, in N. Handa and T. Ohsumi (eds.), *Direct Ocean Disposal of Carbon Dioxide*, Terrapub, pp. 183–193.

Nakicenovic, N. and Swart, R. (eds.) (2000) *Emissions Scenarios. 2000 Special Report of the Intergovernmental Panel on Climate Change*, Cambridge University Press. ISBN 92 9169 113 5.

Needham, J. (1959) *Science and Civilisation in China, Volume 3, Mathematics and the Sciences of the Heavens and the Earth*, Cambridge University Press, p. 465.

Nemani, R.R., Keeling, C.D., Hashimoto, H. *et al.* (2003) Climate driven increases in global terrestrial net primary production from 1982 to 1999. *Science*, **300**, 1560–1563.

Newell, R.G. and Stavins, R.N. (2000) Climate change and forest sinks: factors affecting the costs of carbon sequestration. *Journal of Environmental Economics and Management*, **40**, 211–235.

Nordhaus, W.D. (1994) *Managing the Commons: The Economics of Climate Change*. MIT Press. ISBN 0–262–14055–1.

Nordhaus, W.D. (2007) Critical assumptions in the Stern Review on climate change. *Science*, **317**, 201–202.

Nordhaus, W.D. and Boyer, J. (1999) *Warming the World: Economic Models of Global Warming*, Yale University.

Nordhaus, W.D. and Boyer, J. (2000) *Roll the DICE Again: Economic Models of Global Warming*, Yale University.

Ormerod, B. (1997) *Ocean Storage of Carbon Dioxide: Workshop 4 – Practical and Experimental Approaches*. IEA Greenhouse R&D Programme. ISBN 1 898373 05 1.

Orr, J.C. and Aumont, O. (1999) Exploring the capacity of the ocean to retain artificially sequestered CO_2, in P. Riemer, B. Eliasson and A. Wokaun (eds.), *Greenhouse Gas Control Technologies*, Elsevier, pp. 281–286.

Orr, J.C. and Sarmiento J.L. (1992) Potential of marine macroalgae as a sink for CO_2: constraints from a 3-D general circulation model of the global ocean. *Water, Air, and Soil Pollution*, **64**, 405–421.

Ortiz, R., Sayre, K.D., Govaerts, B., *et al.* (2008) Climate change: can wheat beat the heat? *Agriculture, Ecosystems and Environment*, **126**, 46–58.

Ozaki, M. (1997) CO_2 injection and dispersion in mid-ocean depth by moving ship. *Waste Management*, **17**, 369–373.

Pan, D., Guan, W., Ban, Y. and Huang, H. (2005) Ocean primary productivity estimation of the China Sea by remote sensing. *Progress in Natural Science*, **15**, 627–632.

Peng, T.-H. and Broecker, W.S. (1991) Dynamical limitations on the Antarctic iron fertilisation strategy. *Nature*, **349**, 227–229.

Rau, G.H. and Caldeira, K. (1999) Enhancing carbonate dissolution: a means of sequestering waste CO_2 as ocean bicarbonate. *Energy Conversion and Management*, **40**, 1803–1813.

Redfield, A.C., Ketchum, B.H. and Richards, F.A. (1963) The influence of organisms on the composition of sea water, in M.N. Hill (ed.), *The Sea, Volume 2, Wiley Interscience*, pp. 26–77.

Revelle, R. and Suess, H.E. (1957) Carbon dioxide exchange between atmosphere and ocean and the question of an increase of atmospheric CO_2 during the past decades. *Tellus*, **9**, 18–27.

Richards, K.R., Moulton, R.J. and Birdsey, R.A. (1993) Cost of creating carbon sinks in the USA. *Energy Conservation and Management*, **34**, 905–912.

Riemer, P.W.F. (1996) Greenhouse gas mitigation technologies, an overview of the CO_2 capture and storage and the future activities of the IEA Greenhouse Gas R&D Programme. *Energy Conversion and Management*, **37**, 665–671.

Ritschard, R.L. (1992) Marine algae as a CO2 sink. *Water, Air and Soil Pollution*, **64**, 289–303.

Royal Society (2005) *Ocean Acidification Due to Increasing Atmospheric Carbon Dioxide*, Policy document 12/05, The Royal Society (www.royalsoc.ac.uk).

Rusin, N. and Flit, L. (1960) *Man Versus Climate*. Peace Publishers.

Salter, S., Sortino, G. and Latham, J. (2008) Sea-going hardware for the cloud albedo method of reversing global warming. *Philosophical Transactions of The Royal Society A*, **366**, 3989–4006.

Sato, T. and Sato, K. (2002) Numerical prediction of dilution process and biological impacts in CO_2 ocean sequestration, *Journal of Marine Science and Technology*, **6**, 169–180.

Sato, T., Jung, R.-T. and Sato, K. (2002) Numerical simulation of CO_2 droplet behaviour, in *Proceedings of the 5th International Symposium on CO_2 Fixation and Efficient Utilisation of Energy, 4th International World Energy System Conference, Tokyo, 2001*, pp. 48–53.

Seifritz, W. (1990) CO_2 disposal by means of silicates. *Nature*, **345**, 486.

Shindo, Y., Fujioka, Y., Yanagisawa, Y., Hakuta, T. and Komiyama, H. (1995) Formation and stability of CO_2 hydrate, in N. Handa and T. Ohsumi (eds.), *Direct Ocean Disposal of Carbon Dioxide*, Terrapub, pp. 217–231.

Shoji, K. and Jones I.S.F. (2001) The costing of carbon credits from ocean nourishment plants. *The Science of the Total Environment*, **277**, 27–31.

Simmonds, M. and Hurst, P. (2005) Post combustion technologies for CO_2 capture: technoeconomic overview of selected options, in M. Wilson, J. Gale, E.S. Rubin *et al.* (eds.) *Greenhouse Gas Control Technologies: Proceedings of the 7th International Conference on Greenhouse Gas Control Technologies, 5–9 September 2004, Vancouver, Canada*, Elsevier, pp. 1181–1186.

Singarayer, J.S., Ridgewell, A. and Irvine, P. (2009) Assessing the benefits of crop albedo bio-geoengineering. *Environmental Research Letters*, **4**, 045110. DOI: 10.1088/1748-9326/4/4/045110.

Smith, I.M. (1999) *CO_2 Reduction: Prospects for Coal*, IEA Greenhouse Gas R&D Programme. ISBN 92 9029 336 5.

Snowdon, P., Eamus, D., Gibbons, P. *et al.* (2000) Synthesis of allometrics, review of root biomass and design of future woody biomass sampling strategies. NCAS Technical Report No. 17. Australian Greenhouse Office, Canberra.

Stanhill, G. (2007) A perspective on global warming, dimming, and brightening. *EOS, Transations American Geophysical Union*, **88**, 58. DOI: 10.1029/2007EO050007.

Steinberg, M. (1999) Fossil fuel decarbonization technology for mitigating global warming. *International Journal of Hydrogen Energy*, **24**, 771–777.

Stern, N. (2007) *The Economics of Climate Change. The Stern Review*. Cambridge University Press.

Stevens, S.H., Kuuskraa, V.A., Gale, J. and Beecy, D. (2001) CO_2 injection and sequestration in depleted oil and gas fields and deep coal seams: worldwide potential and costs. *Environmental Geosciences*, **8**, 200–209.

Swaminathan, B. and Sukalac, K.E. (2004) Technology transfer and mitigation of climate change: the fertilizer industry perspective. Presented at the IPCC Expert Meeting on Industrial Technology Development, Transfer and Diffusion, Tokyo, Japan, 21–23 Sept. 2004.

Takahashi, T., Sutherland, S.C., Sweeney, C. *et al.* (2002) Global sea-air CO_2 flux based on climatological surface ocean pCO_2, and seasonal biological and temperature effects. *Deep-Sea Research Part II*, **49**, 1601–1622.

Thomas, W.H. (1969) Phytoplankton nutrient enrichment experiments off Baja California and in the eastern equatorial Pacific Ocean. *Journal of the Fisheries Research Board of Canada*, **26**, 1133–1145.

Thomas, W.H. (1970) Effect of ammonium and nitrate concentration on chlorophyll increases in natural tropical pacific phytoplankton populations. *Limnology and Oceanography*, **15**, 386–394.

Thornton, H. and Shirayama, Y. (2001) CO_2 ocean sequestration and its biological impacts, III-1: Effects of CO_2 on benthic organisms. *Nippon Suisan Gakkaishi*, **67**, 756–757 (in Japanese).

Tiessen, H. (ed.) (1995) *Phosphorus in the Global Environment*. John Wiley & Sons.

Tokimatsu, K., Sorai, M., Kaya, Y. *et al.* (2004) Evaluation of benefits of CO_2 ocean sequestration, in M. Wilson, J. Gale, E.S. Rubin *et al.* (eds.) *Greenhouse Gas Control Technologies: Proceedings of the 7th International Conference on Greenhouse Gas Control Technologies, 5–9 September 2004, Vancouver, Canada*, Elsevier, pp. 773–781. http://uregina.ca/ghgt7/PDF/papers/peer/158.pdf.

Tol, R.S.J. (2002) Estimates of the damage costs of climate change – part I: Benchmark estimates. *Environmental and Resource Economics*, **21**, 47–73.

Toyota, T., Nakashima T., Fujita, T. and Ishii, S. (1991) Surface sea water fertilization experiment with deep sea water, in Hirano, T. (ed.), *Study of the Development of an Effective Technique to Utilise Deep Sea Water Resources*. Japanese Science and Technology Agency, Research and Development Bureau, Tokyo (in Japanese), pp. 87–96.

Tranter, D.J. and Newell, B.S. (1963) Enrichment experiments in the Indian Ocean. *Deep-Sea Research*, **10**, 1–9.

Wadsley, M.W. (1995) Thermodynamics of multi-phase equilibria in the CO_2–seawater system, in N. Handa, and T. Ohsumi (eds.), *Direct Ocean Disposal of Carbon Dioxide*, Terra Scientific Publications, pp. 195–216.

Wanninkhof, R. (1992) Relationship between gas exchange and wind speed over the ocean. *Journal of Geophysical Research*, **97**, 7373–7382.

Ware, D.M. and Thomson, R.E. (2005) Bottom-up ecosystem trophic dynamics determine fish production in the northeast Pacific. *Science*, **308**, 1280–1284.

Weart, S.R. (1997) The discovery of the risk of global warming. *Physics Today*, January, 34–40.

Wigley, T.M.L., Richels, R. and Edmunds, J.A. (1996) Economics and environmental choices in the stabilization of atmospheric CO_2 concentration. *Nature*, **379**, 240–243.

Winjum, J.K., Dixon, R.K. and Schroeder, P.E. (1992) Estimating the global potential of forest and agro forest management practices to sequester carbon. *Water, Air and Soil Pollution*, **64**, 213–227.

World Heath Organization (2006) *Health Effects of the Chernobyl Accident and Special Health Care Programmes*, WHO.

Wright, A., Lackner, K. and Peters, E. (2006) World Patent WO 2006/084008.

Index